Advanced Algebra Study Guide

An engineering approach to learning, understanding, and teaching Algebra

A visual, procedural and verbal, learning guide.

Mr. Cook
Principal Electrical Engineer
Adjunct Math Professor
November 13, 2020 [1]

Copyright Page

PCN 9781735463506

Library of Congress Control Number: 2020914441

ISBN: 9781735463506 Advanced Algebra Study Guide

0.1 Appreciation

My Family

Susan S., Library Consultant

Brandon, Math Student from FSU

Professor Walt Jensen, CNEC Math Department

2

Contents

List of Figures

Prefix

0.2 Introduction

After a career in Electrical and Computer Test engineering, then teaching math in college I compiled much of my notes used in the classroom to help future students gain math instinctive problem-solving skills, using the tools presented in class.

0.3 Scope of this Book

The Scope of this book is to provide a study guide for many of the principles covered in my College Algebra Courses.

0.4 Intention for Results of the study guide

The result of this study is intended to give the student a better understanding of a solid approach to solving math Problems. In subsequent tests and in the Engineering field the student may need to re-derive the principle or equation to solve the problem presented by the required task. The student learns to use underlying principles to create mathematical flexibility to solve problems in the class room or engineering tasks in the workplace.

0.5 My Math website: Mathstudyguide.org

Please feel free to check out my website: mathstudyguide.org

Chapter 1

Algebra Overview

The basic mode of algebra is to develop rules for handling unknowns, coefficients and numbers and exponents.

1.1 Letter Variables

The basic letter variables of Algebra are usually $(x,\ y,\ \text{and } z)$ and possibly $(a,\ b,\ \text{and } c)$
Any letter can be used depending on the application required.

1.2 State of variables during an Algebra Problem

The Algebraic variable is constant for the length of the solution.
Note: In calculus the variables are $(v,\ t,\ d, \text{and } a)$ constantly changing during the problem.
For instance, a drag race car if accelerating at a constant rate the velocity is continually changing. The time and distance are changing as well.

1.3 General Characteristics of Algebra

Algebra is a Math that can transform a formula from a word problem which is composed of add terms and subtracted terms into factors of multiplies. When in the multiplied factors format, we can solve the quadratic equation by setting the equation to either y (for the graph) or zero (for the x intercepts or roots).

$x^2 - 4x - 96 = 0$	Solve for x-intersection (solution)
$x^2 - 4x - 96 = y$	Graph on a graphing calculator

$96 = x(x-4)$	Original Equation
$96 = x^2 - 4x$	Original Equation adds and subtracts
$(x-12)(x+8) = 0$	Factored form (factors); can be solved

Refer to page 104 for the solution.

1.4 Cross reference to solving an equation by factoring

Another example of a quadratic problem solution can be found on page 115

Example: Calculate the time the ball takes to hit the ground when thrown upwards from a certain height.

1.5 Zero-Factor property, rule

The Zero-Factor Rule, Explanation is found on page 22

We can use the zero-factor rule.

When there two factors equal to zero then either factor or both are equal to zero.

Therefore, set each factor equal to zero and solve.

$(x+1)(x-7) = 0$ By the zero−factor property set each factor equal to zero

$$(x+1)=0\,; \quad x=-1$$
$$(x-7)=0\,; \quad x=7$$

1.6 Variables in a problem Algebra Problem using the same formula

In a further example the constants representing the variable will change while the formula will be reused for different initial and final conditions. In other words, the formula can be re-used with different constants but using the same variables without changing the equation. In summary in algebra the attribute is that the variable stays the same.

Chapter 2

Real Numbers

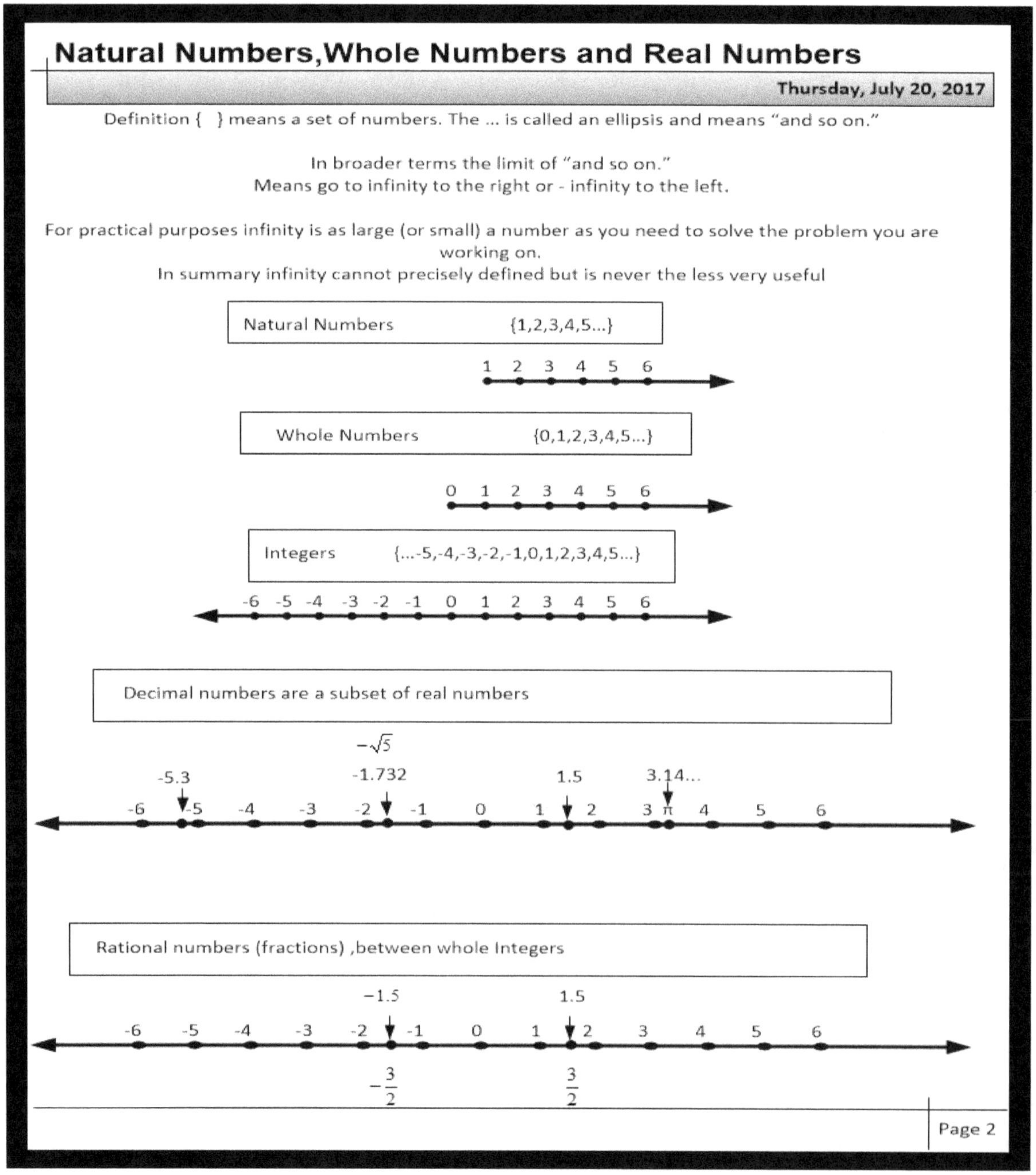

The diagram is an example of the types of real number lines in the real number system. Note that the rational number system consists of all the numbers on the real number line. In a graph of the real number line this would represent the x-axis in the Cartesian Coordinate System .

1. natural numbers
2. whole numbers
3. Integers
4. Rational numbers
5. Irrational numbers
6. Imaginary numbers

2.1 Explanation of number types

1. Natural Numbers Counting numbers, starts at +1 and increases to the right.
2. Whole Numbers, Add a Zero to the left of the +1 of natural numbers
3. Integers, Add -1 to the left of the zero and continue left to as far as needed.
4. Rational numbers, Numbers that can be expressed as a fraction
5. Any terminating decimal is rational.
6. For instance 0.666... where the digit goes on indefinitely is rational.
7. Any repeating decimal pattern is rational
8. For instance .6732732732... where the pattern 732 is repeated indefinitely.
9. Integers are rational numbers since the can be divided by one (1)

10. Irrational numbers cannot be expressed as a fraction $Pi = 3.14159$ is an irrational number.

11. Imaginary numbers are the result of taking the square root of a negative number.

12. $\sqrt{-4} = \pm 2i$ where $i = \sqrt{-1}$

2.2 Exponential numbers are raised to a power greater than zero or one

An example is x^2

Exponential numbers are raised to a power greater than zero or one

x^2 is an exponential term with x as the base and 2 as the exponent.

2.3 Exponential numbers are raised to a power less than zero or a negative exponent

1. An example is y^{-3}
2. This would equal $\frac{1}{y^3}$

Chapter 3

Plotting two coordinate points

3.1 Cartesian Coordinate System

The Cartesian coordinate system consists of the real number line plotted on the x-axis and the y axis. The y axis is perpendicular the x-axis. The two lines intersect at x, y coordinates at point 0,0

The diagram following is an example of the types of real number lines in the real number system. Note that real number system consists of all the numbers on the real number line.

In a graph of the real number line this would represent the x-axis.

The Cartesian coordinate system consists of the real number line plotted on the x-axis and the y axis. The y axis is perpendicular the x-axis.

3.2 Calculate the slope

The two lines intersect, perpendicularly at $\{\mathrm{x}, \mathrm{y}\} = \{0, 0\}$

Slope Problem: Plot (1.-2) and (-4,7) on the Cartesian coordinate system graph.

Plot the x coordinate first then the y-coordinate. (x,y) The x-coordinate is first in the ordered pair of points. (x, y)

Connect the two points and plot the slope.

$$\text{m} = \text{slope} = \frac{\text{change in y}}{\text{change in x}} = \frac{\Delta\text{y}}{\Delta\text{x}} = \frac{\text{rise}}{\text{run}}$$

$$\text{m} = \frac{9}{-5} = 1.8$$

3.3 Graph of plotting two points

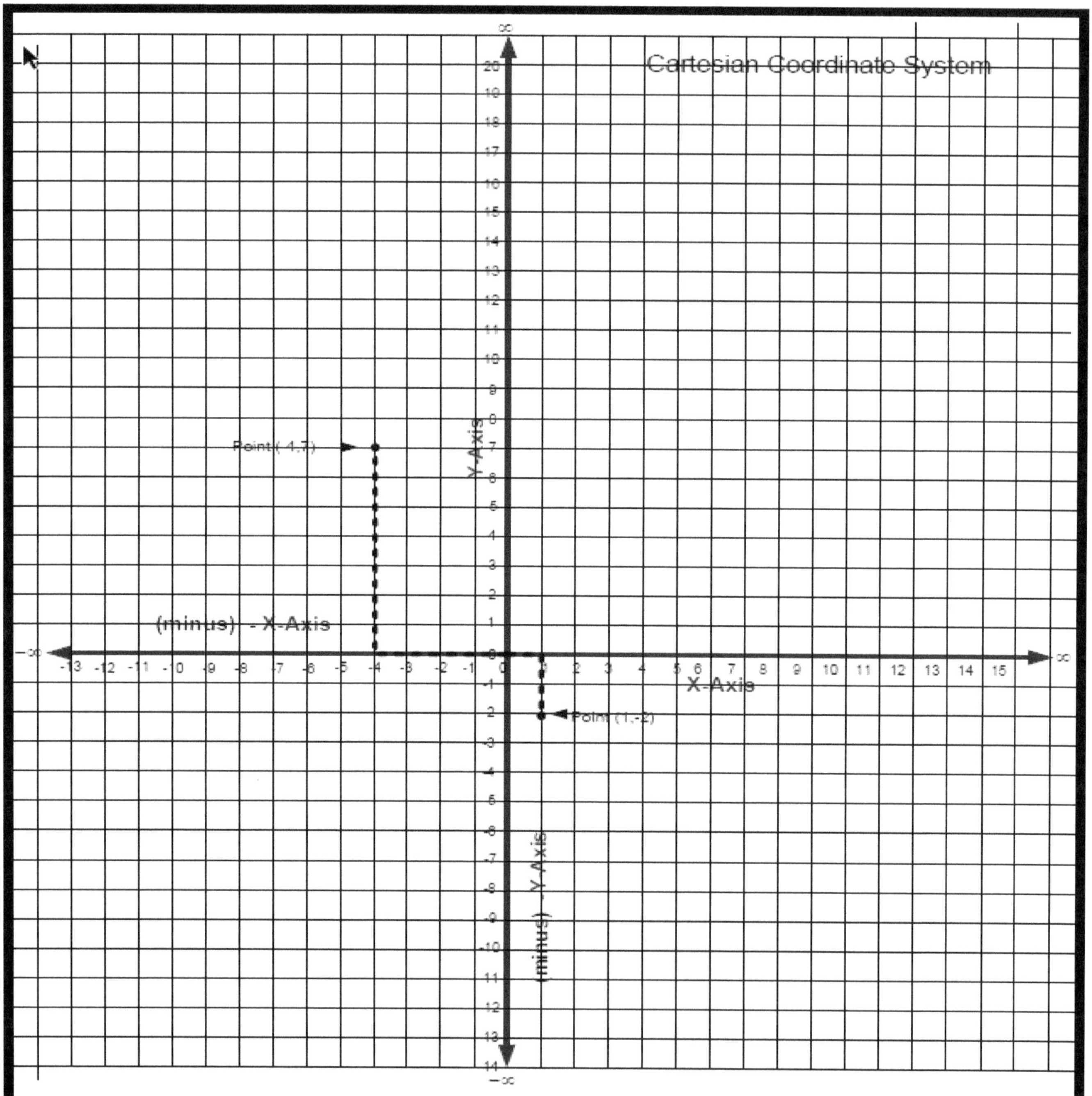

Problem: Plot (1.-2) and (-4,7) on the cartesian coordinate system graph. Plot the x coordinate first then the y-coordinate. (x,y),where the x-coordinate is first in the ordered pair of points.

Figure 3.1: Graph of plotting two points

3.4 Graph of plotting connecting the two points

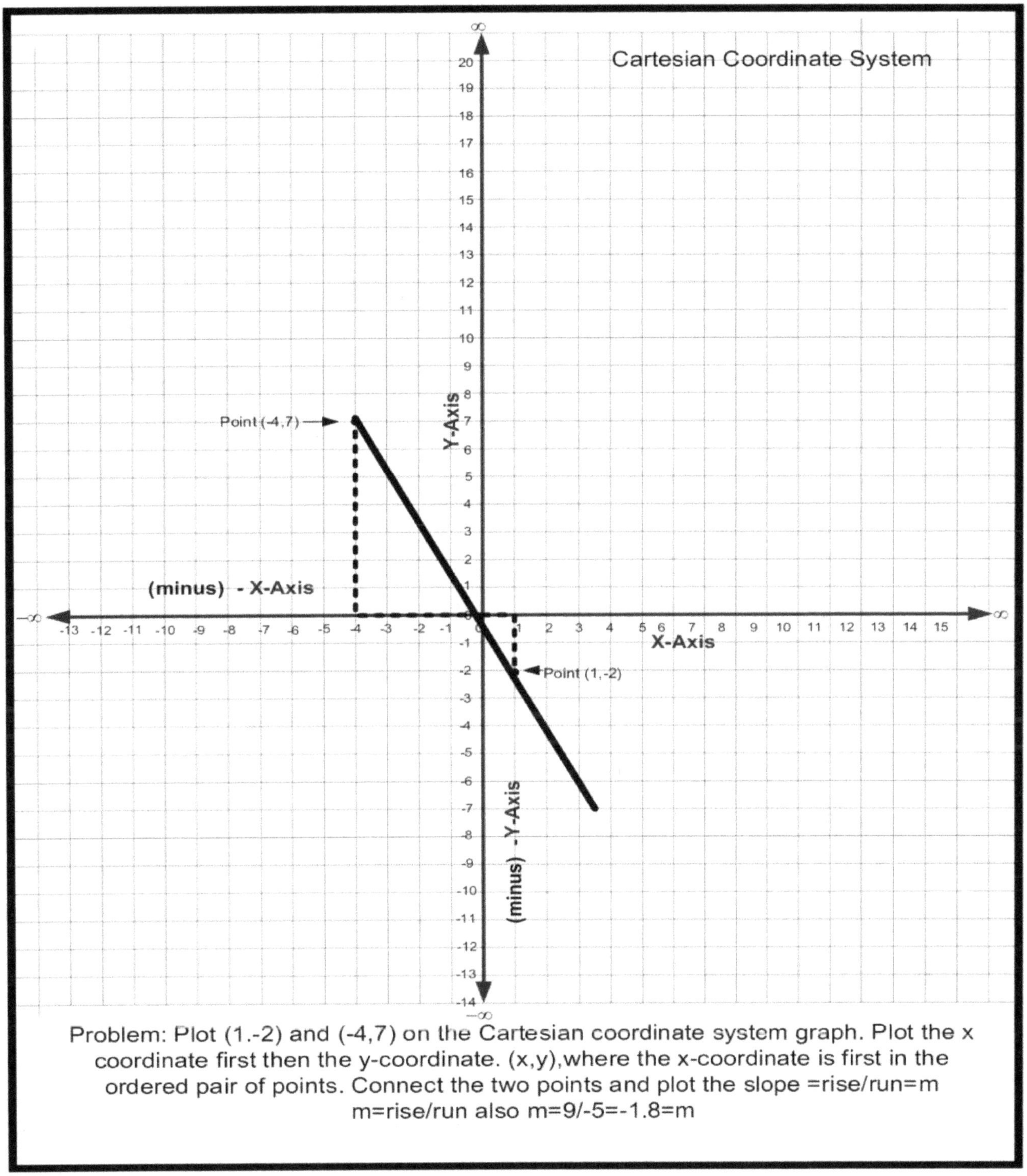

Figure 3.2: Graph of plotting connecting the two points

Chapter 4

Graphs

4.1 Functionality of graphs in the problem-solving process

A basic graph plotting 2 points can be seen on page 11.

In Mathematics graphs are extremely useful to see how a curve behaves. Points such as ($x-$ intersect) and ($y-$ intersect) can usually be seen (given observing that section of the curve) also, as well as the shape of the curve.

In algebra a curve can be a straight line, (Linear) as well as a curved line.

The shape of the curve can be a straight line, a parabola, a sine wave, a tangent wave, etc.

In electronics the curve exhibited by a diode,has a distinctive characteristic curve which is used in designing a circuit.

4.2 Plotting Equations

The Object of plotting equations is to observe the behavior of the equations.

The solution can be often observed, by graphing the equation and observing intersect points and slopes.

Chapter 5

Rules for factoring expressions

5.1 Rules for factoring

Properties of Real Numbers	**General Algebraic Formula**	**Real Numbers Examples**	**Properties used to solve expressions and equations**
Closure Property Closure Property	a + b is a unique number a * b is a unique number	12 + 6 =18 12 * 6 =72	These solutions have only one real value Note the results of a square root has two solutions
Communicative Property of Addition	a + b = b + a	4 + x = x + 4	The order of operation of addition does not affect the solution
Communicative Property of Multiplication	ab = ba	12 * 6 = 6 * 12	The order of operation of Multiplication does not affect the solution
Associative Property of Addition	(a + b) +c = a + (b + c)	(12 + 6) + 5 = 12 + (6 + 5)	Grouping (Associating) in Addition does not affect the solution
Associative Property of Multiplication	(ab)c = a(bc)	(3*7)5 =3(7*5)	Grouping (Associating) in Multiplication does not affect the solution

5.2 More Rules for factoring expressions

Properties of Real Numbers	**General Algebraic Formula**	**Real Numbers Examples**	**Properties are used to solve algebraic expressions**
Identity Property a + 0 0 + a a * 1 1 * a	These are special case numbers with unique properties	a + 0 = a 0 + a = a a * 1 = a 1 * a =a	Adding 0 to a does not change its value Multiplying 1 to a does not change its value
Inverse Property of Addition Using Additive Inverse	Property of opposites For each(**a**) there is a unique real number (**- a**) such that a + (-a) = 0 (**-a**) is the addative inverse of(**a**)	15 + (-15) = 0 and (-15) + 15 = 0 x - 20 = 0 x-20+(20)=0+20 x = 20 solution	This rule is used to clear side of an equation by adding the additive inverse to both sides

5.3 Even more rules for factoring expressions

Properties of Real Numbers	**General Algebraic Formula**	**Real Numbers Examples**	**These Properties are used to solve expressions and equations**
Distributive Property of Multiplication	a(b+c) =ab+ac and (b+c)a = ba+ac or ab+ac	12a*(6b+5c) =72ab+60ac and (6b+5c)*12a =72ab+60ac	The law of distribution is used to multiply two algebraic expressions such as a monomial times a binomial
Inverse Property of Multiplication (using the multiplicative inverse.)	For each **a** except **zero** there is a unique number (a) such that the reciprocal is $(\frac{1}{a})$ (a)*$(\frac{1}{a})$=1 or $(\frac{1}{a})$*(a)=1)	$(20)*\frac{1}{20} = 1$ $\frac{1}{20} * (20) = 1$	Multiplying a term by its reciprocal results in **1** This will resolve a coefficient. of a single term.

Chapter 6

Factoring

6.1 Factoring

These Algebra rules can be found on page 16.

Algebra has sets of rules to manipulate and solve algebraic examples or problems, by working with unknown values, their coefficients and their exponents.

Algebraic equations can be viewed as graphs to show the overall characteristic of the equation.

6.2 What is factoring?

Factors are the result of Multiplication A simple example would be 5 times x equals $5x$ Therefore both 5 and x are factors of x.

$$5x = 5 * x$$

where $5 * x$ are factors of $5x$
Another example is $40x^2$

$$40x^2 = 2 * 2 * 2 * 5 * x * x$$

Then $2 * 2 * 2 * 5 * x * x$ are factors of $40x^2$

6.3 Zero factor property process

Anything multiplied by 0 is 0.

Basically, if $A * B = 0$ then either A *or* B or both can be zero.

Based on this theory we can set both factors equal to zero and solve.

Example of a solution of both factors set to 0.

$(x+1)(x-7)=0$

$$(x+1)=0\,; \quad x=-1$$
$$(x-7)=0\,; \quad x=7$$

6.4 Factoring from a process (overview) stand point

- Factor the equation into factors.
- Each factor is multiplied by the remaining factors.
- Solve the remaining factors by setting each factor to zero and solve for each factor.

Algebraic Rules for reducing algebra expressions

Factoring uses rules to reduce an Algebraic expression with add and subtracts operators into an equivalent where the expressions consist of are multiply and divide operators.

These rules are found on page 16 Algebra Rules for factoring.

How do we use factoring in Algebra?

When an algebraic expression is divided up into factors, we can use the zero-factor property to solve the expression.

6.5 Characteristics of a solving quadratic equation

6.5.1 A quadratic can be set to y or y=0

Note that if an equation equal zero we are graphically solving the point intersects with the x-axis.

For instance, if $ax^2 + bx + c = 0$ in its general form we can solve for the equation for x .

If $ax^2 + bx + c = y$ or $y = ax^2 + bx + c = y$ then we are considering the equation and we can plot the equation on a TI-73 or equivalent calculator or on a program like geometers sketchpad.

6.6 An example of a specific trinomial equation

$x^2 + 4x + 4 = 0$ when factored $= (x-2)(x-2) = 0$

By setting $x - 2 = 0, x = 2$ and $x - 2 = 0, x = 2)$

The solutions are $x = 2, 2$ therefore both solutions are $x = 2$ when $y = 0$ so the parabola described by $x^2 + 4x + 4 = 0$ contacts the x-axis only one time.

6.7 Conclusion

The equation graphs as $y = x^2+4x+4$ When we set $y = 0$ then $0 = x^2+4x+4$ or $x^2 + 4x + 4 = 0$. By definition when the value of $y = 0$ and a value of x shows where the curve touches or crosses the x or -x axis. **This graph can be found on 164,Review and Graph of quadratics**

6.8 Factoring Methods

Factoring has rules that are followed to solving an algebraic problem. **These rules are found on page 16**

6.9 Factoring definition

In general in algebra we have 2 types of equations

- Equation with adds operators and subtract operators.
- These equations need to be converted to factors.
- Equation with multiply operators and divide operators.
- Equations with multiply factors can be solved by setting each factor equal to zero.

6.10 Multiplying and Factoring

Multiplying and Factoring are opposite operations.

- If we multiply two algebraic expressions together such a $3x * 2y = 6xy$
- We can factor $6xy$ into its component parts we get $6xy = 2 * 3 * x * y$
- if we combine 2 pairs of factors,we get, either $2x * 3y$ or $3x * 2y$
- $2x * 3y = 6xy$ and $3x * 2y = 6xy$
- Therefore, we can multiply $3x * 2y = 6xy$
- Then we can factor $6xy = 2 * 3 * x * y$ into its component parts.
- To check our answer, we multiply the component parts and get the same answer of $6xy$.

6.11 Mulitplying Binomials,Using FOIL

- FOIL is a method of multiplying 2 binomials.
- $(x + 5)(x + 8)$
- The F (first terms) multiplied are the $x * x = x^2$

- The O(outer terms) are $x * 8 = 8x$
- The I(Inner terms) are $5 * x = 5x$
- The L(Last terms) are $5 * 8 = 40$
- Adding up the terms including two like terms

$$x^2 + 5x + 8x + 40$$

$$x^2 + 13x + 40$$

6.12 Trinomial factoring

1. The purpose of the following diagram is to show in programming diagram format the decisions made to solve this trinomial into binomial factors
2. The rounded rectangle shows the start of the process.
3. The diamond represents a decision to be made in solving the problem.
4. The Rectangle shows the result of the solution.
5. The shaded in shapes show the path of the decision for this example
6. The flow diagram is shown in order to develop a more logical approach to factoring by examining the middle term and the sign of the third term.

6.13 By Deriving the Middle Term of a trinomial, we define the trinomial to binomial factors process

A trinomial can (if factor-able) can be factored into 2 binomial factors

This changes the format from adds and subtracts into two multiplied factors.

When the factors are set to zero use the Zero Factor Property to solve.

by setting each factor to Zero and solving

The Zero Factor Explanation is found on page 22

$ax^2 \pm bx \pm c = (x \pm n)(x \pm m)$

Multiply binomial terms

$$(x+n)(x+m) = x^2 + nx + mx + nm$$

6.14 Reverse the multiplication process to factor

Collect middle terms

$$(x+n)(x+m) = x^2 + (n+m)x + nm$$

The constants $\boldsymbol{n}$ and $\boldsymbol{m}$ are added as coefficients

to the second or middle term of the trinomial

The constants $\boldsymbol{n}$ and $\boldsymbol{m}$ are multiplied

for the constant in the third term.

Reverse the process to factor since "factoring undoes Multiplication"

Factor $\boldsymbol{n}\ \boldsymbol{m}$ and add the terms to get the

correct middle term by "trial and error" process.

Chapter 7

Factor by Grouping

7.1 Summary for Factor by Grouping

When solve quadratic equations with the coefficient of the first term, or the squared term we:

Factor out any monomial term common to each term in the quadratic equation.

1. In the case of

$$-24x^2 - 42x + 45 \text{ we can factor out } -3$$

$$\text{giving us } -3(8x^2 + 14x - 15)$$

2. Note that we factored a -3 instead of a $+3$ giving us a positive coefficient for the first term.

3. If the coefficient of the squared term still does not equal one then factor by grouping.

4. We need to factor $(8x^2 + 14x - 15)$. Since the coefficient of the x^2 is 8 we factor by grouping.

7.2 Steps for factor by grouping

1. Solve
$$-24x^2 - 42x + 45 \text{ we can factor out } -3$$
$$\text{giving us} -3(8x^2 + 14x - 15)$$
2. Multiply the coefficient of the first term times the third term integer,
3. Split the middle term coefficient $+14$ into two terms that are equivalent to the middle term
4. Group into 2 binomial terms
5. Factor any common binomial
6. 3x is the largest factor that can be factored out of all 3 terms. Note: We divide out -3x instead of +3x because we need to have the first term inside the parenthesis be a positive value.
7. Factored out binomial -3
$$-3(8x^2 + 14x - 15)$$

7.3 Split the middle term into 2 equivalent terms

1. In a quadratic equation we multiply two terms to get the third term.
2. Factoring is the opposite of multiplying so the third term cam be split into factors.
3. Multiply the first term coefficient time the third term integer,
4. Multiplying
$$8 \bullet (-15) = -120$$
5. In a quadratic equation we add two terms to get the coefficient of the middle term.
6. The factors of the third term need to add up to the middle term.

7.4 Factor -120

1. The easiest way to factor is to divide the number by a small prime number, such as 2, 3, or 5,sometimes 7 or a higher prime number.

2. nm = -120 If we start by dividing by 2 and continue dividing by 2 we get

$$-1 \bullet 2 \bullet 2 \bullet 2 \bullet 15$$

3. Since we have run out of $2's$ we divide by the next prime number 3

$$-1 \bullet 2 \bullet 2 \bullet 2 \bullet 3 \bullet 5$$

7.5 Combine the factors into two terms that add up to $+14$ the middle term

1. factors

$$-1 \bullet 2 \bullet 2 \bullet 2 \bullet 3 \bullet 5$$

2. Test

$$8 + (-15) = -7 \quad \text{Fails}$$

3. Test

$$-6 + 20 = 14 \qquad \text{Passes}$$

4. Since

$$-6 + 20 = 14 \quad \text{Middle term and } (-6 \bullet 20) = -120 \text{ Third term}$$

5. This is the correct combination of factors

7.6 Rewrite the middle term from 14 to $-6 + 20$ which is an equivalent term

1. Now we have 2 factors that add up the middle term we can rewrite the equation: (Don't forget to save the -3x factor for the final list of

factors)

$$-3(8x^2 + 14x - 15) \ = -3(8x^2 + 20x - 6x - 15)$$

2. Rewrite the middle term

$$-3(8x^2 - 6x + 20x - 15)$$

3. Group two sets of terms into factors

$$-3(8x^2 - 6x)(+20x - 15)$$

4. Factor out monomials.

$$-3 \quad ((2x)\,(4x - 3) + ((5)\,(4x - 3))$$

5. Factor out binomial $(4x - 3)$

$$-3 \bullet ((4x - 3) \bullet (2x + 5))$$

7.7 Check by using FOIL to multiply

1. FOIL multiply

$$-3 \bullet \left(8x^2 + 20x - 6x - 15\right)$$

$$-3 \bullet \left(8x^2 + 14x - 15\right)$$

2. Multiply

$$-3 \bullet \left(8x^2 + 14x - 15\right)$$

3. The calculation checks out.

$$-24x^2 - 42x + 45 \quad = -24x^2 - 42x + 45 \quad \text{Original equation}$$

7.8 Quadratic Equation from the quadratic logic diagram.

Factor by grouping is a more direct solution to this problem.

$$12x^2 - x - 20$$

Multiple the first term coefficient and the third term constant.

$$12 * (-20) = -240$$

$$\text{Factor}[-240] = (-1)\left(2^4\right)(3)(5) = (-16)(15)$$

Combine the factors that add up to the coefficient of the middle term.

$$-16 \text{ and } 15 \text{ add up to } -1$$

which is the correct coefficient of the middle term.

now we rewrite the middle term resulting in 4 terms.

$$12x^2 - x - 20$$

becomes

$$12x^2 - 16x + 15x - 20$$

Note : $-16x + 15x$ is equivalent to $-x$ when added Now Group the terms and factor.

$$\left(12x^2 - 16x\right) + (15x - 20)$$

$$4x(3x - 4) + 5(3x - 4)$$

Now factor out the $(3x - 4)$ resulting in

$$(3x - 4)(4x + 5) \quad \text{Answer}$$

Chapter 8

Factoring using the Difference of squares method

8.1 Difference of squares is a Special Factoring product.

$(a+b)(a-b) = a^2 - b^2$

$(a+b)(a-b) = a^2 + ba - ba - b^2$

Since the middle term drops out when you multiply the two binomials.

therefore $a^2 - b^2 = (a+b)(a-b)$

This product is called **The Difference of Squares** product.

Examples of difference of squares

$$x^2 - 9 = (x-3)(x+3)$$

$$x^4 - 1 = \left(x^2 - 1\right)\left(x^2 + 1\right)$$

Intermediate step

$$\text{Note } \left(x^2 - 1\right) \text{ is also a perfect square}$$

Therefore:

$$x^4 - 1 = (x - 1)(x + 1)\left(x^2 + 1\right)$$

Another example of Differences of squares

$$x^2 - 4y^2 = (x + 2y)(x - 2y)$$

8.2 Difference of squares is a Special Factoring product.

Examples of difference of squares

$$x^2 - 9 = (x - 3)(x + 3)$$

$$x^4 - 1 = \left(x^2 - 1\right)\left(x^2 + 1\right)$$

Intermediate step

$$\text{Note } \left(x^2 - 1\right) \text{ is also a perfect square}$$

Therefore:

$$x^4 - 1 = (x - 1)(x + 1)\left(x^2 + 1\right)$$

Another example of Differences of squares

$$x^2 - 4y^2 = (x + 2y)(x - 2y)$$

Chapter 9

Special product factoring

9.1 Perfect Square Trinomials

The special products rule for an algebraic term ,squared.

$$(a+b)^2 = a^2 + 2ab + b^2$$
$$(a-b)^2 = a^2 - 2ab + b^2$$

9.2 Perfect Square Trinomial

A perfect square trinomial is the square of a binomial

$x^2 + 10x + 25$ is a perfect square trinomial.

25 is a perfect square.

the middle term test: 2 times the product of the square root of the first and third term.

Note: $\sqrt{x^2} = x \quad \sqrt{25} = 5$

$$2 * x * 5x = 10x$$

The middle term is correct.

The middle term test passes

Therefore, the binomial consists of the addition of the two square roots squared.

$(x+5)^2$ *answer*

9.3 Perfect Square Trinomial

$x^2 - 22x + +121$ is a perfect square trinomial.

121 is a perfect square.

the middle term test: 2 times the product of the square root of the first and third term.

The result of a square root $\sqrt{121} = -11$ or 11
Note : Two solutions

Note: $\sqrt{x^2} = x \quad \sqrt{121} = 11$

Note: $-\sqrt{x^2} = -x \quad \sqrt{121} = -11$

$$2 * x * -11x = -22x$$

The middle term is correct.

The middle term test passes

Therefore, the binomial consists of the subtraction of the two square roots squared.

$(x-11)^2$ *answer*

Chapter 10

Factoring decisions logic flow chart

The method to solve this equation intuitively look at several factors:

Is the coefficient of the first term one?

If the coefficient greater than one? Do factor by grouping.

Is the middle term coefficient large? Then one of the m's or n's will be large,

Are the signs of the second term and third positive?

Is the coefficient of the third term negative? The factored terms will have opposite signs Plus-Minus or Minus-Plus

If the third term coefficient is positive both signs of the binomials will be the same. Plus-Plus or Minus-Minus

Chapter 11

Small Middle term

$ax^2 \pm bx \pm c = (x \pm n)(x \pm m)$
$12x^2 - x - 20$

The Coefficient of the middle term is small compared to the coefficients of the first term and the constant in the third term. Therefore, choose small terms in binomials and coefficients will be close in value.

Sign of c last term positive? No

Sign of b middle term positive? No

Sign factors are opposite where $m < n$

See logic chart for highlighted decision path,

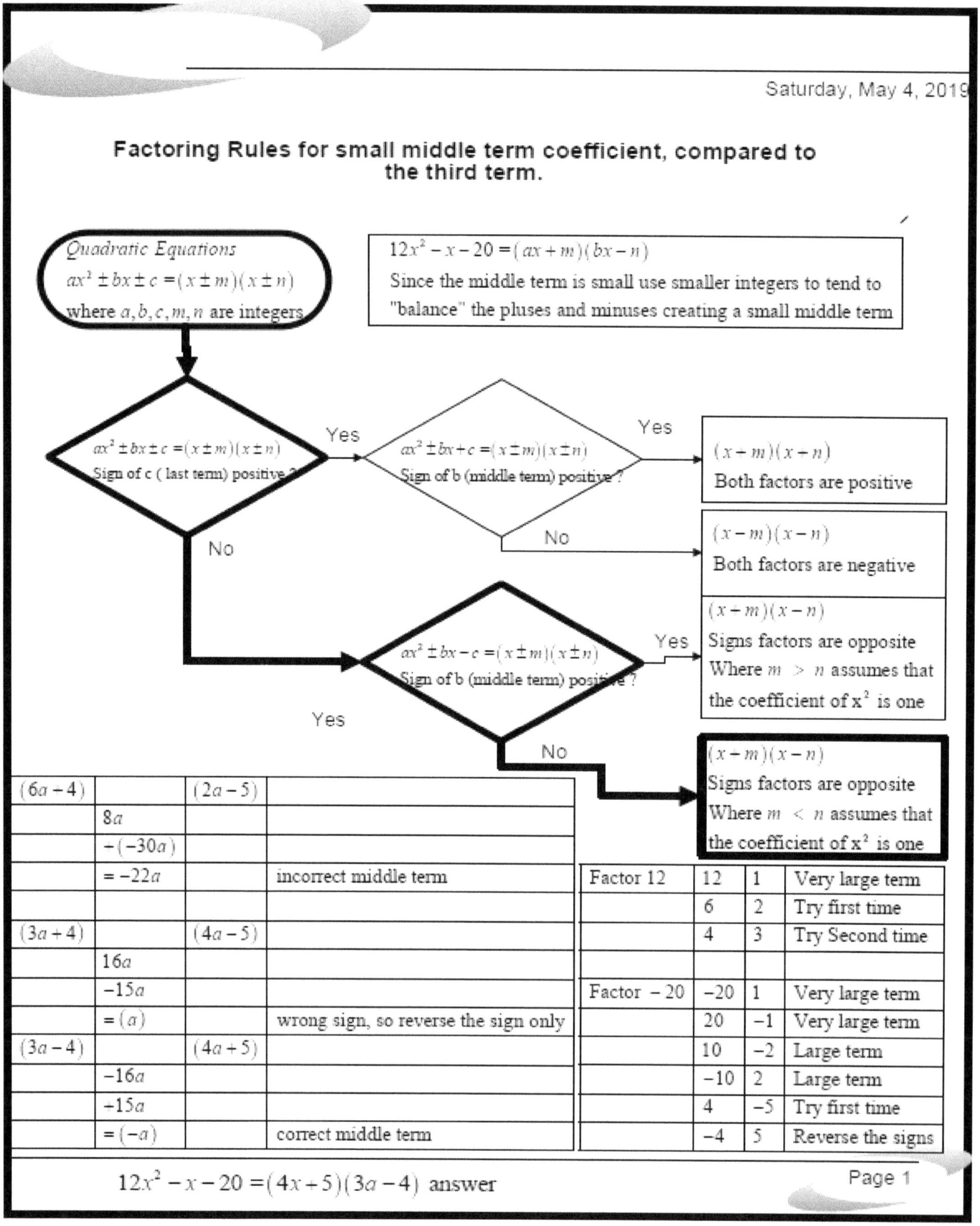

Figure 11.1: Factoring Logic Chart

11.1 Factor common factor, then factor quadratic expression

Diagrams of factoring rules

Tuesday, September 8, 2020

$4x^2+12x-40$ =a$(x\pm m)(x\pm n)$ first factor out the common factor 4

$4(x^2+3x-10)$

1) Note: if there is a negative sign with the third term the signs must be different since the third term is a result of multiplying two factors together and $(m)(-n)=-mn$ also $(-m)(n)=-mn$

so in general terms $4(x^2+3x-10)=4(x\pm m)(x\mp n)$

2) The two integers n and m needed to be added together to get the value +3 the coefficient of x.

3) Some of the factors of -10 are $[-5,2]$ and $[5,-2]$

try $4(x-5)(x+2)$. The middle term is $-5+2=-3$

4) Wrong sign only so switch the signs $4(x+5)(x-2)$ which gives $5-2=3$.*Correct* coefficienbt of the middle term.

$4x^2+12x-40$ =$4(x+5)(x-2)$ Check by FOIL

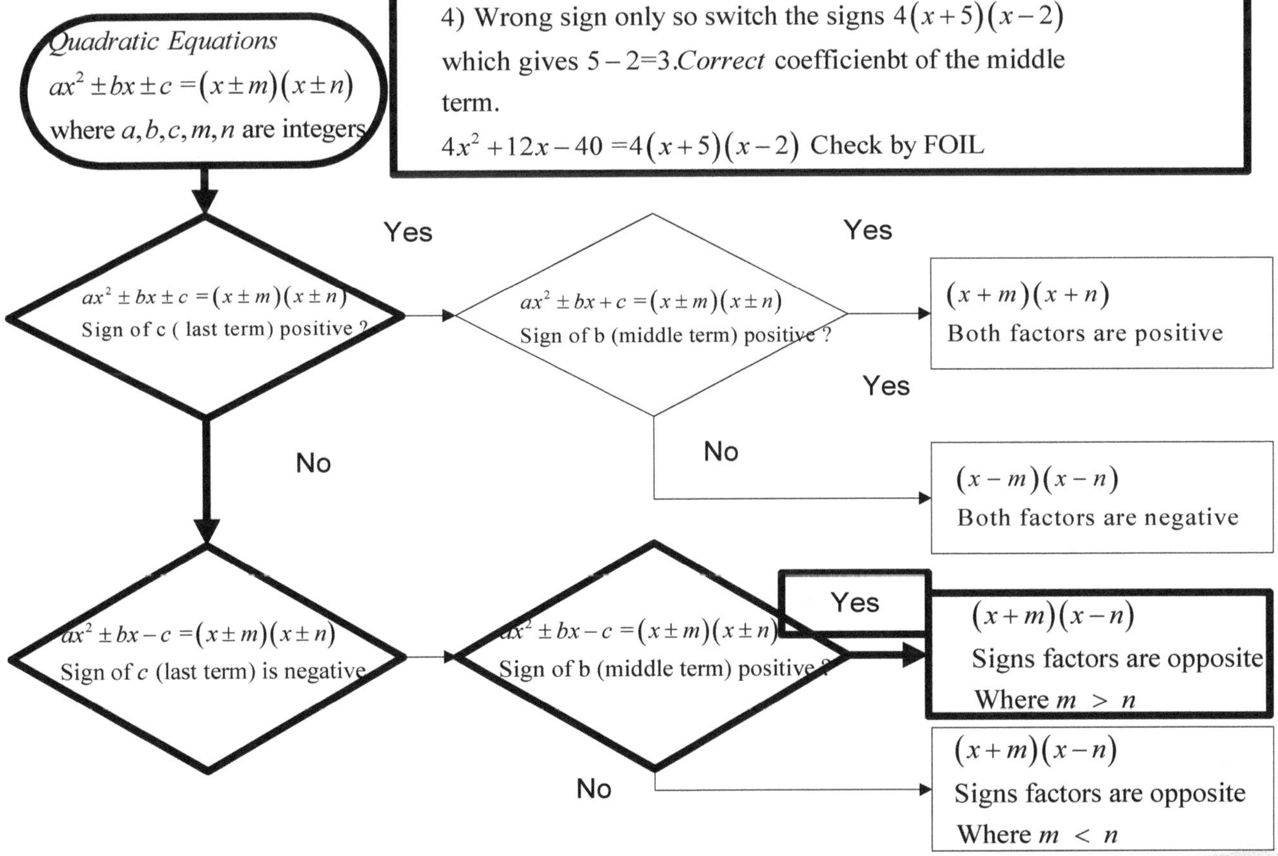

Page 1

Chapter 12

Factoring examples

12.1 Examples

The basic mode of algebra is to develop rules for handling unknowns coefficients and numbers and exponents. The advantage in learning algebra is that the same algebraic rules apply to trigonometry and calculus Trig and calculus layer on more characteristics on algebra.

12.2 Letter Variables

The basic letter variables of Algebra are usually $x.y.z$ and a, b, c Any letter can be used depending on the application required

12.3 Factor out any monomial term common to each term in the quadratic equation. Then solve the remaining trinomial

For example, $3x^2 + 12x + 12$. The common factor is 3.
Factoring gives us $3(x^2 + 4x + 4)$.

This can be factored to $3(x+2)(x+2)$ or $3(x+2)^2$ using factoring techniques.

Chapter 13

Polynomials Basic Algebraic Forms

13.1 Polynomials

Polynomials are basically in 3 separate formats using a combination of numbers and letters raised to some power.
Examples:

$3x^2$	Monomial
$(x+5)$	Binomial
x^2+4x+4	Trinomial
$x^4+2x^3+3x^2+5x+7$	Polynomial

$x^4+2x^3+3x^2+5x+7$
has a number of terms greater than 3 terms and is simply named a polynomial.
In general all the expressions list above are also call polynomials.

13.2 Monomials

A monomial ,mono meaning one, consists of a letter representing a variable

$$x, y, z, a, v, t, a, b, c$$

and a number preceding the variable letter variable does not show, and the one is not necessary, then the number preceding it is assumed to be one. In general, the 1 exponent, is not specified but assumed, therefore, the one is not necessary to express the x. Therefore $x = 1x$

Examples of monomials are,

$$x, 3x^2, 10y, \frac{z}{3}$$

13.3 Binomials

Binomials (Bi meaning two) are expressions of two monomials separated by a plus (+) or minus (-) operators, and enclosed by parenthesis.

13.4 Examples of binomials

$(x - 3)$ is separated by a minus sign and is one entire unit designated as a binomial. $(3x + 5)$ is another example of a binomial.
Note the parenthesis around a binomial. In general, a binomial is $(ax \pm b)$

13.5 Trinomials

Trinomials (tri for three) means 3 terms. The usual trinomial form we see in algebra is of the form $ax^2 + bx + c$ it may in many cases be factored into two binomial factors i.e. $(x \pm a)(x \pm b)$ where x is the variable and a and b are constants.

A typical example of a trinomial is $x^2 + 4x + 4$ where a=1, b=4, and c=4.

Much of learning algebra in the early part of study is learning the rules and methods of factoring. **These rules are on page 16. Chapter 6**

13.6 Trinomials continued

More advanced applications come later in the learning and involve understanding and applications of algebra.

Note that algebra is used in Physics, Chemistry, Trigonometry and Calculus.

The algebraic rules do not change. The main difference is that functions will be applied along with variables, such as sin, cos, tan, csc, sec and cot, in trigonometry.

In calculus the variables are changing during the solution to the problem, whereas the variable does not change during the solution in Algebra.

Also the variables in Physics v (velocity) a (acceleration) t(time) and (distance) and have specific physical properties.

13.7 Polynomials with an exponent of 4 or higher

For example

$$x^4 + 3x^2 + 2x + 15$$

This Polynomial may be factorable.
In the case it is not factorable it is said to be in its prime state.

Chapter 14

Linear equations, three solution types including the graphing of the equations.

14.1 2 types of solutions to solving a pair of linear equations

- In order to better understand the characteristics of linear equations , we can solve them as
- Solving by addition, Check by Graphing the two equations.
- Solving by Cramer's rule.
- Check by Graphing the two equations.

In engineering we solve equations by different methods to verify the solution.

Chapter 15

Example Solving simple linear equations 1 unknown variable x.

The following pages show solutions for linear equations. Linear are solved for the variable x.
Since y=f(x)
we substitute the value of x into the original
equation and solve for the value of y giving us the coordinates of an intersection given 2 linear equations.

15.1 Solving Linear Equations one variable x Example 1

Solve:

$$2x + 3 = 17$$

$$2x + 3 - 3 = 17 - 3$$

add -3 to both sides to get isolate the $2x$ Use the additive inverse -3

$$2x = 14$$

$$\frac{2x}{2} = \frac{14}{2}$$

Divide both sides by 2 to isolate the x Use the multiplicative inverse $\frac{1}{2}$

$$x = 7 \quad \text{Solution.}$$

15.2 Check your answer by substituting the value of in the original equation

$$2x + 3 = 17$$

$$2(7) + 3 = 17$$

$$17 = 17$$

$$\text{The solution } (17) \text{ checks out}$$

15.3 Solving Linear Equations 1 variable x, 2 equal equations Example 2

Solving Linear Equations Example

$$2x - 17 + x = 3x + 1 + 2x$$

$$3x - 17 = 5x + 1$$

Solve each side then isolate the x by combining both sides.

$$3x - 17 - 5x = 5x + 1 - 5x$$

$$-2x - 17 = 1$$

$$-2x - 17 + 17 = 1 + 17$$

$$-2x = 18$$

$$\frac{-2x}{2} = \frac{18}{2}$$

$$-x = 9$$

$$x = -9$$

Solution $x = -9$
Check by substituting in the original equation.

$$2x - 17 + x = 3x + 1 + 2x$$

$$2(-9) - 17 + (-9) = 3(-9) + 1 + 2(-9)$$
$$-18 - 17 - 9 = -27 + 1 - 18$$
$$-44 = -44$$

it checks out !!

The intersection was calculated at

$$\{x, y\} = \{-9, -44\}$$

To find the y-intersection set x=0 and calculate:

$$\{x, y\} = \{0, 1\}$$

and

$$\{x, y\} = \{0, -17\}$$

Chapter 16

Method for solving a two linear equations system

16.1 Solving by addition

Solve the following linear system.

$$\begin{aligned} 3x + 11y &= -9 \\ 7x + 6y &= 38 \end{aligned}$$

$$\begin{vmatrix} 3x + 11y = -9 \\ 7x + 6y = 38 \end{vmatrix}$$

Multiply the first equation by 7 Multiply the second equation by -3 The $x's$ drop out when you add the two equations.

$$\begin{aligned} (3x + 11y = -9) \ \bullet 7 \\ (7x + 6y = 38) \ \bullet -3 \end{aligned}$$

$$\begin{aligned} (3x + 11y = -9) \ \bullet 7 = (21x + 77y = -63) \\ (7x + 6y = 38) \ \bullet -3 = (-21x - 18y = -144) \end{aligned}$$

Add the two equations

$$\begin{aligned} &21x + 77y = -63 \\ &-21x - 18y = -144 \\ &------------ \end{aligned}$$

$$-59y = 177$$
$$y = -3$$

Substitute the value of y in either equation and solve for x.

$$3x + 11(-3) = -93x - 33 = -9$$
$$3x = 24$$
$$x = 8$$

The solution is (x, y)=(8,-3)

Check

$3x + 11y = \ -9$

$3(8) + 11(-3) \ = \ -9$

$24 - 33 = -9$

$-9 = -9$ It checks !!

16.2 Graphical Solution

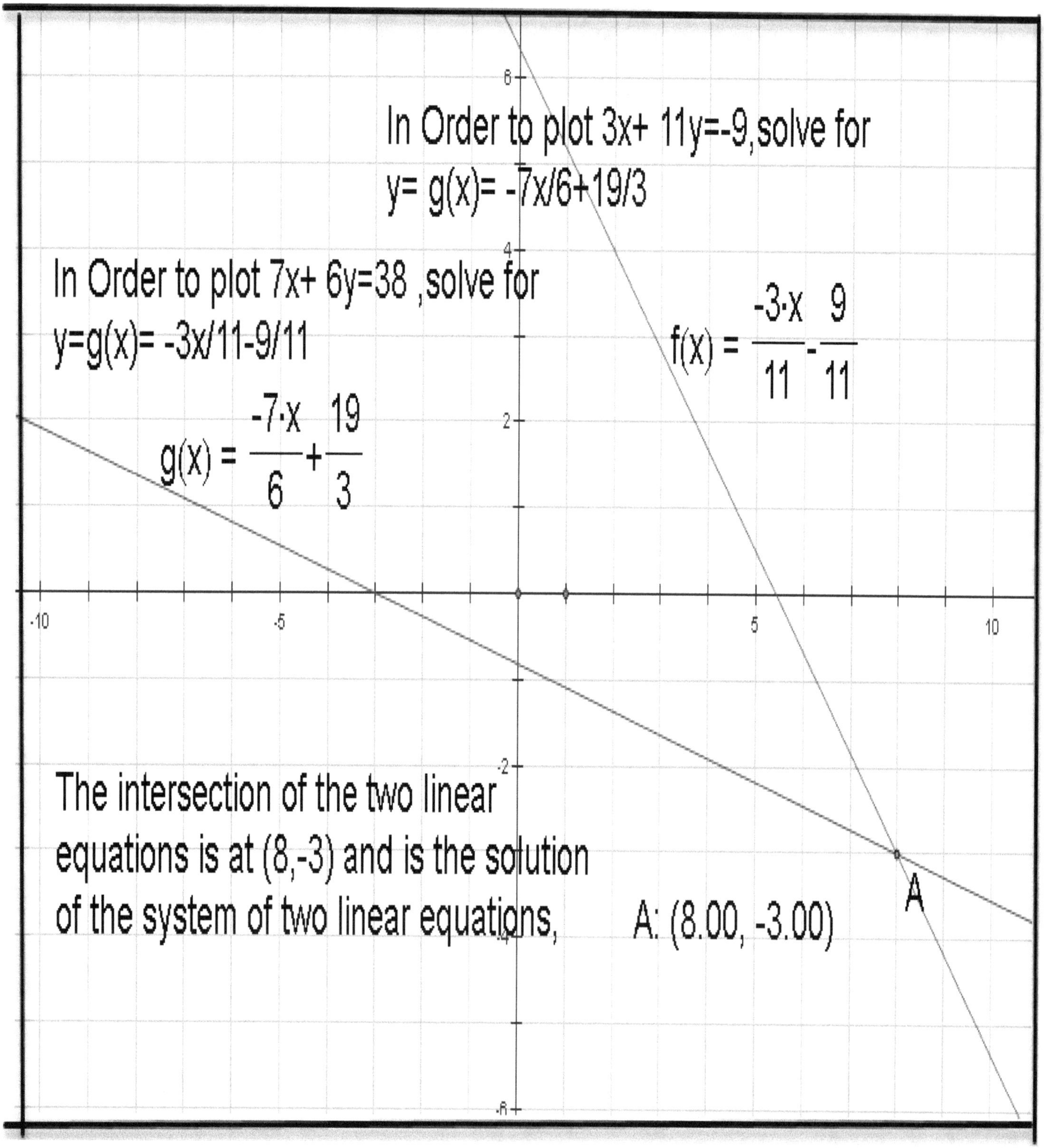

Figure 16.1: Graphical Solution for intersecting linear equations.

Chapter 17

Linear equations

Solutions for linear equations.

$$4(2x + 1) - 29 = 3(2x - 5)$$

In this case the two equations Intersect since their slopes are not equal.

$$y_1 = 4(2x + 1) - 29$$

$$y_2 = 3(2x - 5)$$

17.1 Slopes of left and right side of the equation.

The lines intersect at

$$y_1 = y_2$$

Left side $y_1 = mx + b$ where m is the slope

$$4(2x + 1) - 29$$

reduces to

$$8x + 4 - 29 = 8x - 25$$

where the slope is 8

Right side $y_2 = mx + b$ where m is the slope

$$3(2x - 5)$$

reduces to

$$6x - 15$$

where the slope is 6 Therefore the 2 lines intersect, since the slopes would have to be equal in the case of parallel lines or overlapping lines. If the slopes are not equal the lines would have to intersect at some point,

17.2 Identify two other line types

Note: if the slopes were equal, they would be parallel or overlap.

The lines would be parallel if slope is the same and the $b = y - intercept$ were different. Note if the y-intercepts were the same and the slopes were equal then the lines would overlap.

17.3 Linear equation set example, solving for x, y, and y-intercepts and graphing the solutions

$$4(2x + 1) - 29 = 3(2x - 5)$$

Distribute the coefficient over the binomial in the parenthesis to clear the parenthesis from the equation.

$$8x + 4 - 29 = 6x - 15$$

$$8x - 25 = 6x - 15$$

$$8x - 25 - 6x = 6x - 6x - 15$$

Collect all the x terms on the left side of the equation and add the terms.

$$2x - 25 = -15$$

Collect all the *digits* terms on the right side of the equation and add the terms.

$$2x - 25 + 25 = -15 + 25$$

$$2x = 10$$

$$x = 5$$

Solution $x = 5$ Next we solve for y by substituting the value of $x = 5$ in either side of the original equations. Note that $y = f(x)$ means that the value of y depends on the value of x. We have the value of y from the solution for the equation therefore:

$$4(2x + 1) - 29 = 3(2x - 5)$$

Solving the left side of the equation:

$$y = 4(2(5) + 1) - 29 = 44 - 29 = 15; y = 15$$

Solving the right side of the equation:

$$y = 3(2x - 5)$$

$$y = 3(10 - 5) = 15; y = 15$$

This make sense since at

$$(x, y) = (5, 15)$$

The equations are equal Since the lines intersect this

$$(5, 15)$$

is a unique solution.

17.4 Graphing solution.

We demonstrate this by graphing the lines represented by each side of the equation and observing that the intersection is at

$$(x, y) = coordinate\ points\ (5, 15)$$

17.5 Plot y-intercepts.

Another calculation we make is plotting the y-intersect. By definition the y-intersect is at x=0. Therefore, we solve the left side of the equation by substituting 0 for any x term.

$$y = 4(2x + 1) - 29$$

$$y = 4(0 + 1) - 29 => y = -25$$

for the left side of the equation

also, substitute 0 for any x term on the right side of the equation

$$y = 3(2x - 5)$$

$$y = 3(0 - 5) => y = -15$$

The y intersecting points are

$$(x, y) = (0, -25)$$

for one side of the equation and

$$(x, y) = (0, -15)$$

for the other side of the equation.

[1]

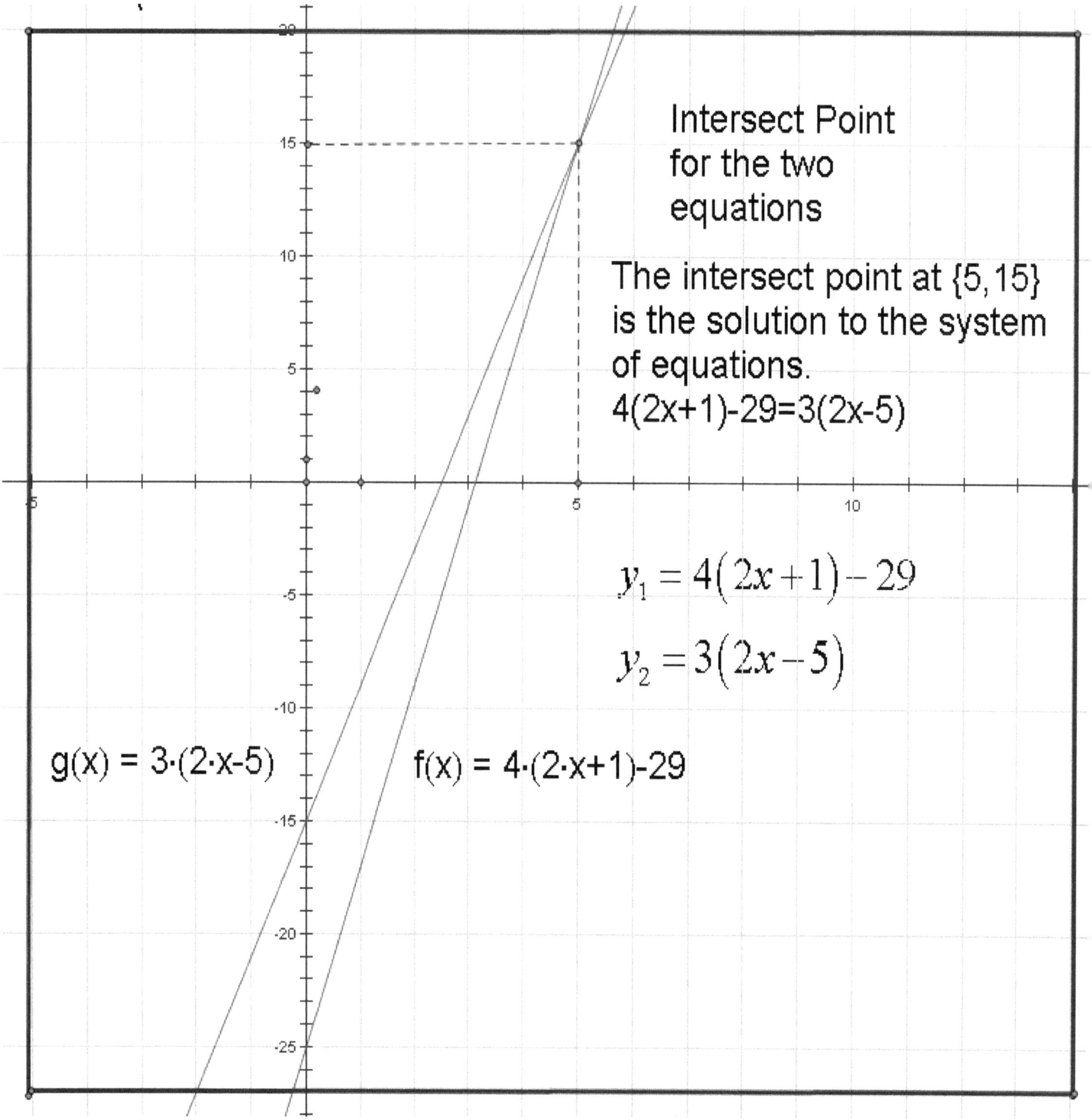

Figure 17.1: Intersect of 2 linear equations

Chapter 18

Another linear equation example.

The graph can be found on page 68

Solving Linear Equations Example

$$2x - 17 + x = 3x + 1 + 2x$$

$$3x - 17 = 5x + 1$$

Solve each side separately

$$3x - 17 - 5x = 5x + 1 - 5x$$

$$-2x - 17 = 1$$

$$-2x - 17 + 17 = 1 + 17$$

$$-2x = 18$$

$$\frac{-2x}{2} = \frac{18}{2}$$

$$-x = 9$$

$$x = -9$$

Solution $x = -9$

18.1 Solve for y_1

$$y_1 = 2x - 17 + x$$

$$y_1 = 2(-9) - 17 + (-9)$$

$$y_1 = -44$$

18.2 Solve for y_2

$$y_2 = 3x + 1 + 2x$$

$$y_2 = 3\,(-9) + 1 + 2\,(-9)$$

$$y_2 = -44$$

$$2(-9) - 17 + (-9) = 3(-9) + 1 + 2(-9)$$

18.3 Solve for Both sides of the equation

Check by substituting in the original equation.

$$2x - 17 + x = 3x + 1 + 2x$$

$$2(-9) - 17 + (-9) = 3(-9) + 1 + 2(-9)$$

$$-18 - 17 - 9 = -27 + 1 - 18$$

$$-44 = -44$$

it checks out !! Also substituting the value for x (-9) yields (-44) in both sided of the equation then $y = (-44)$ and the intersecting point is $\{-9, -44\}$

18.4 Solve for y-intersects

The graph can be found on page 68

To find the y-intersection set x=0 and calculate:

$$\{x, y\} = \{0, 1\}$$

and

$$\{x, y\} = \{0, -17\}$$

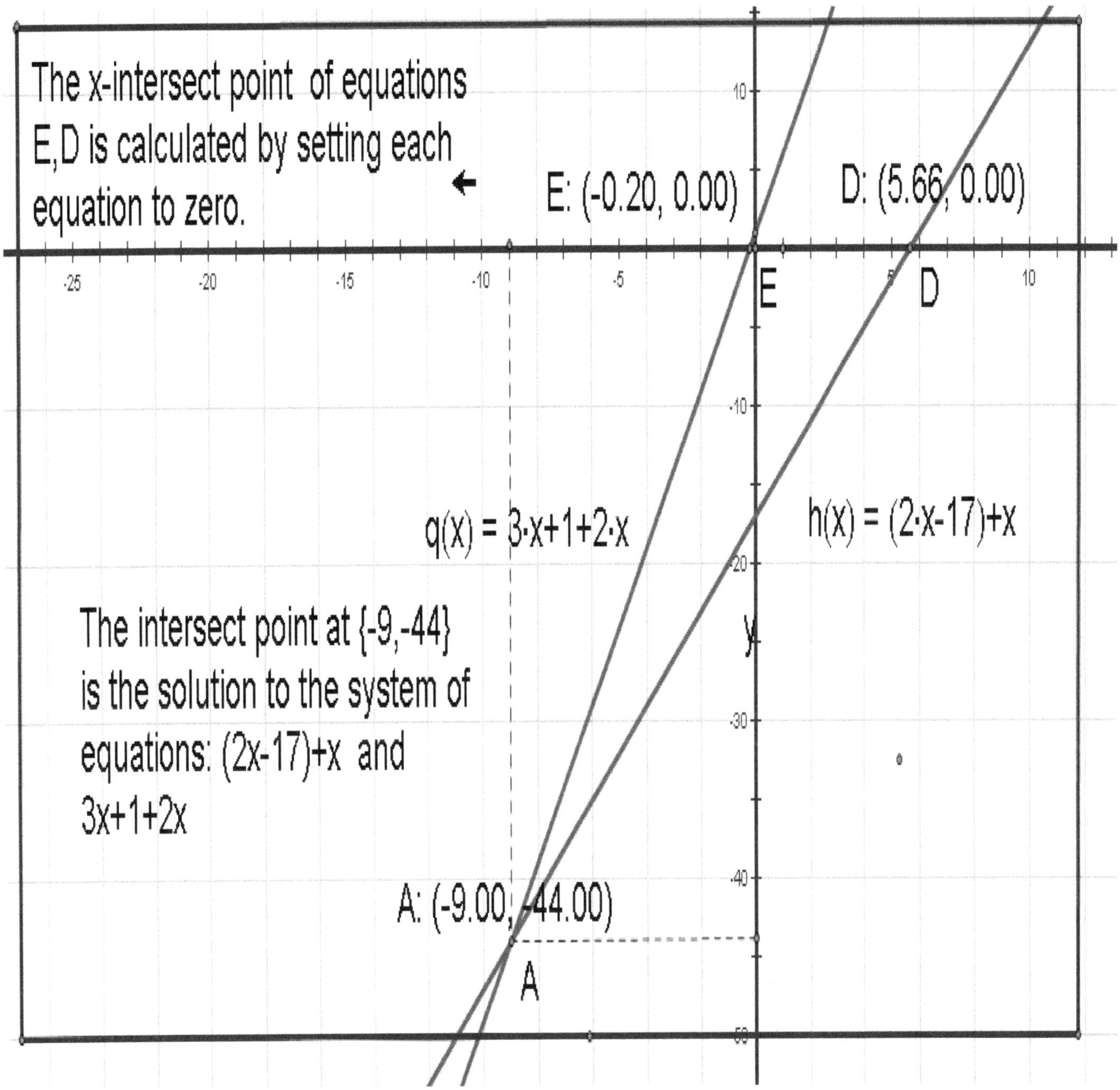

Figure 18.1: Another example of system of 2 equations

Chapter 19

Three possible solutions for system of two Linear equations

When we have two linear equations the solutions can be one of 3 solutions.

- Intersect
- Parallel
- Overlap

A linear equations system has the format: $y = mx + b$

Where the value of y is the dependent variable on x for its value.

x is the independent variable.

b is the y intercept at $x = 0$ creating a point $0, b$

m is the slope and equals

$$\frac{rise}{run} = \frac{\Delta y}{\Delta x} = \frac{vertical\ change\ in\ y}{horizontal\ change\ in\ x}$$

- Intersect $m_1 = m_2$ at one point only; $b_1 \neq \ b_2$ different $\mathrm{y-intercepts}$
- Parallel $m_1 = m_2$ at every point $b_1 = \ b_2$ different $\mathrm{y-intercept}$
- Overlap $m_1 = m_2$ at every point; $b_1 = b_2$ same $\ \mathrm{y-intercept}$

Chapter 20

Three Linear types

Type of Equations	**Equation**	**x, y or Remainder**	**Number of Solutions**	**Lines appear as:**
Consistent (Intersecting)	$y = -x - 1$ $4x - 3y = 24$ See Graph	$\{3, -4\}$	One solution	One Intersecting Point
Inconsistent (Parallel)	$3y - 2y = 6$ $6x - 4y = 24$	x dropped out y dropped out Remainder 0=6 False	No Solution	Lines are Parallel
Dependent (Overlapping)	$y = 3x - 2$ $15x - 5y = 10$	x dropped out y dropped out Remainder 0=0 True	Infinite Number of Solutions	Lines are Overlapping

Figure 20.1: Three Linear types

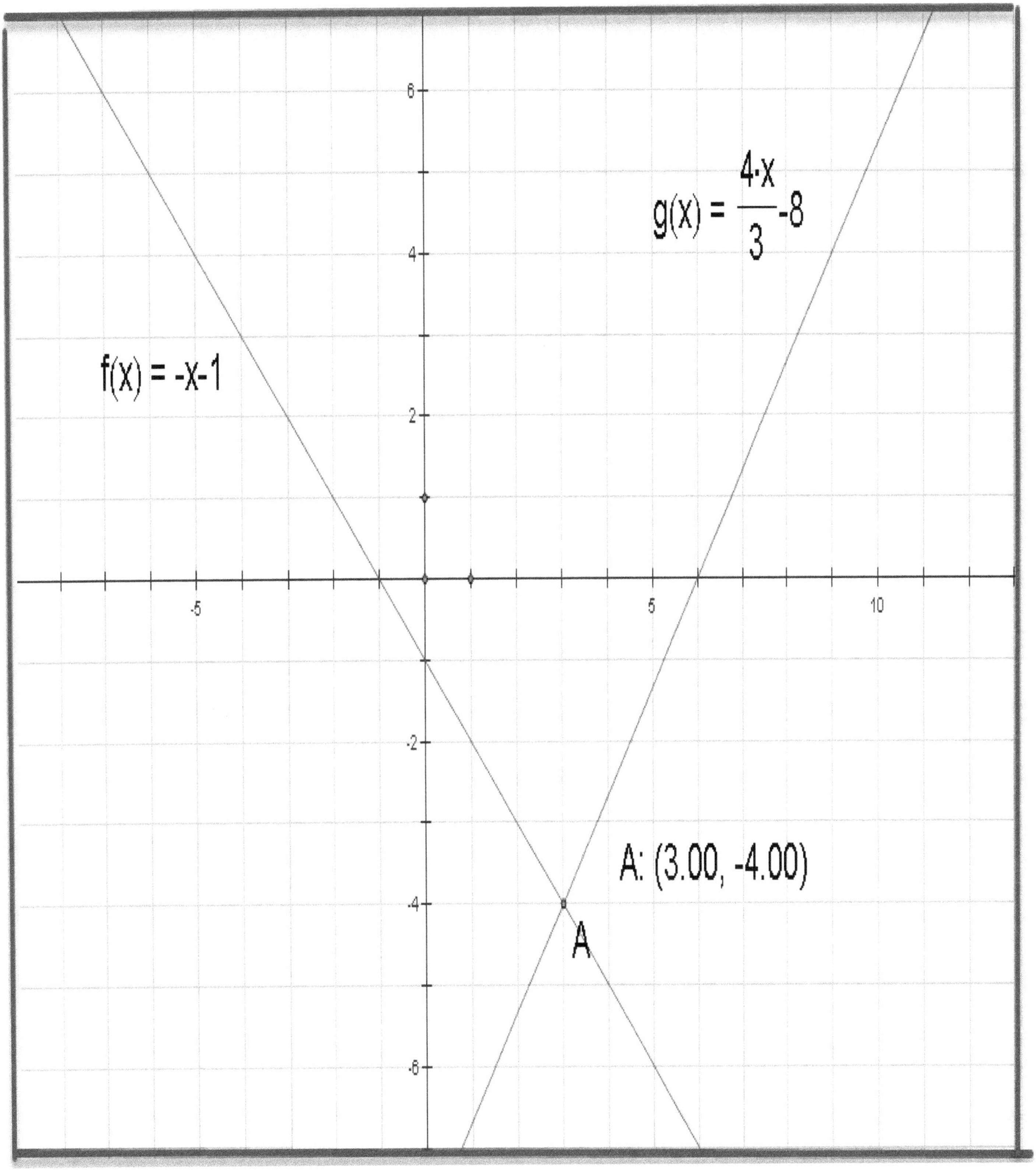

Figure 20.2: Intersecting lines in a linear system

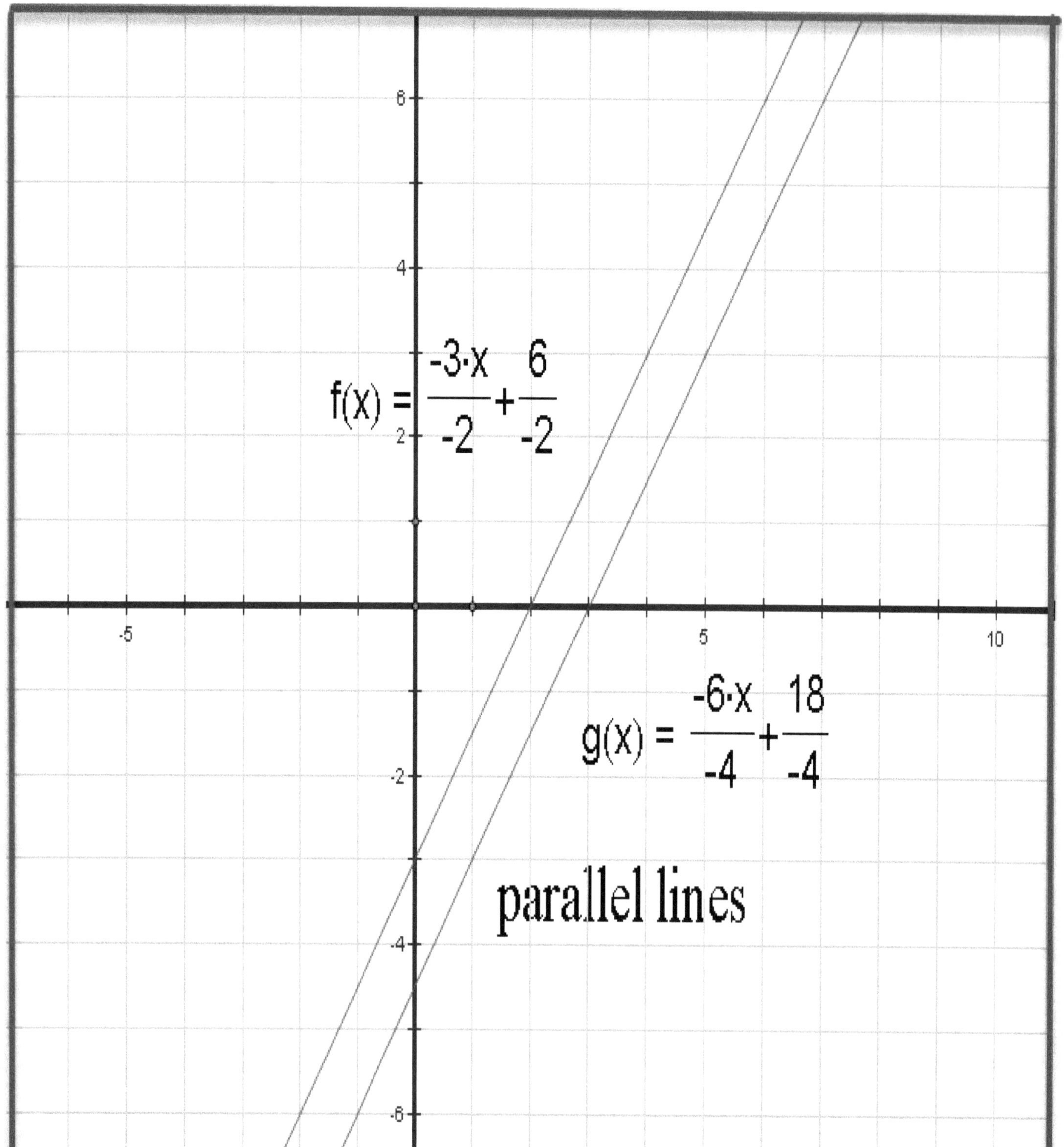

Figure 20.3: Parallel lines in a linear system

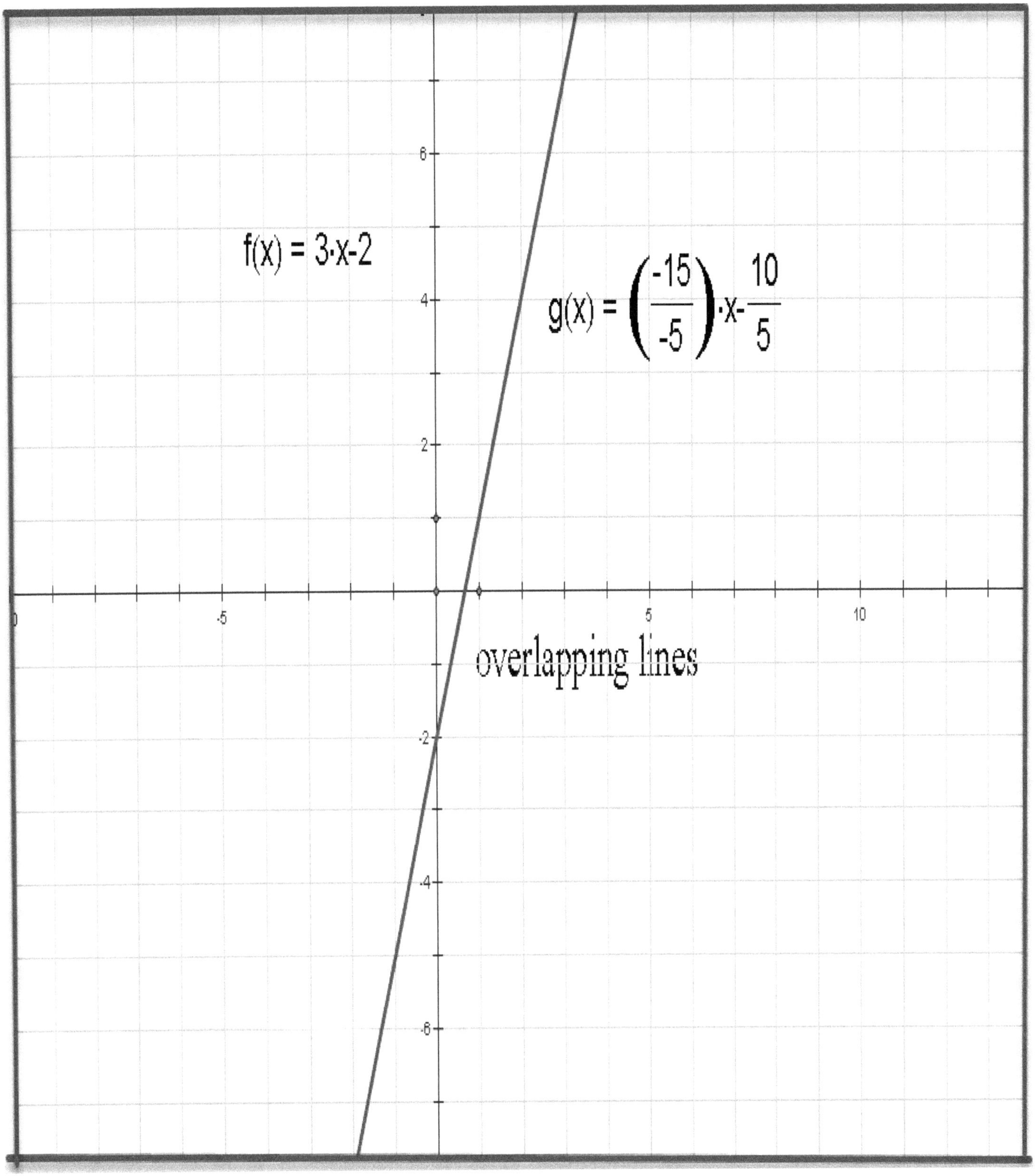

Figure 20.4: Overlapping lines in a linear system

Chapter 21

Solution to 3 linear types

Consistent System

$$y = -x - 1$$

$$4x - 3y = 24$$

Rearrange and solve for y

$$y = -x - 1$$

$$-3y = -4x + 24$$

Multiply the first equation by 3, then add the two equations.

$$3y = -3x - 3$$

$$-3y = -4x + 24$$

$$0 = -7x + 21$$

$$7x = 21$$

$$x = 3$$

$$y = -x - 1$$

$$y = -(3) - 1$$

$$y = -4$$

$$\text{Intersecting Point} = \{3, -4\}$$

Inconsistent System

$$3x - 2y = 6$$
$$6x - 4y = 18$$

$$3x - 2y = 6 \text{ Multiply by } -2$$

$$-6x + 4y = -12$$
$$6x - 4y = 18$$

$$0x - 0y = 6$$
$$0 = 6 \quad \text{False}$$
$$m_1 = m_2 : \text{The slopes are equal}$$

and the lines are parallel

The two lines do not intersect

like railroad tracks. (Hopefully)

Dependent System

$$y = 3x - 2$$
$$15x - 5y = 10$$

$$-3x + y = -2 \text{ Multiply by } 5$$

$$-15x + 5y = -10$$
$$15x - 5y = 10$$

$$0 = 0 \quad \text{True}$$

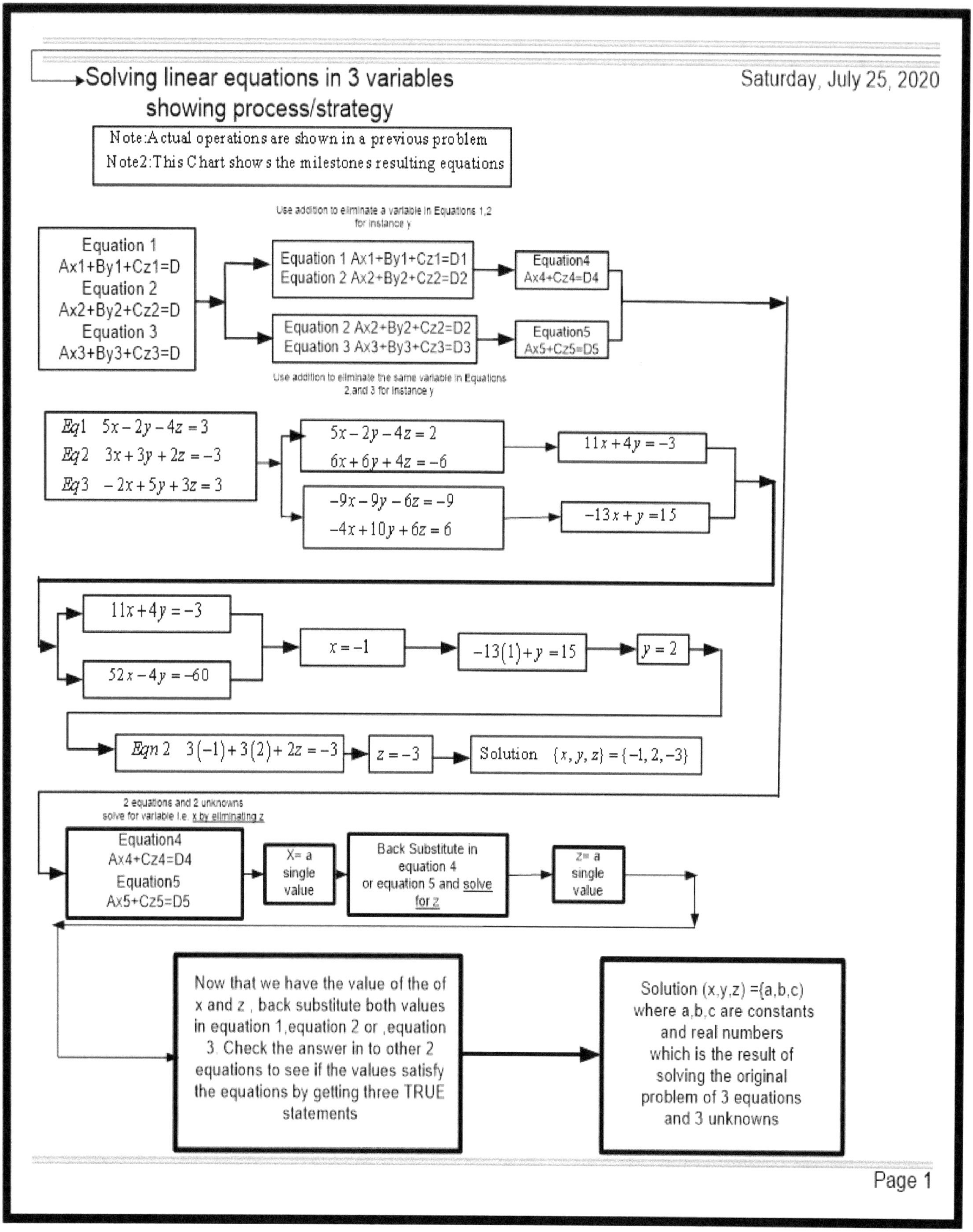

Figure 21.1: Solving 3 variables x 3 unknowns' graph of process-continued

Chapter 22

3 equations and 3 unknowns, elimination by addition

22.1 General Format for 3 equations and 3 variables

General Format

$$
\begin{aligned}
Ax_1 + By_1 + Cz_1 = D_1 \\
Ax_2 + By_2 + Cz_2 = D_2 \\
Ax_3 + By_3 + Cz_3 = D_3
\end{aligned}
$$

22.2 System to be solved 3 variables 3 equations

22.2.1 Reduce the system to 2 equations and 2 variables.

Method using simultaneous equations,

22.2.2 Eliminate variable z

More detail

Equation 1: $5x - 2y - 4z = 3$

Equation 2: $3x + 3y + 2z = -3$

Equation 3: $-2x + 5y + 3z = 3$

Reduce the system to 2 equations and 2 variables.

Equation 1: $5x - 2y - 4z = 3$ No change

Equation 2: $3x + 3y + 2z = -3$ Multiply by 2

Equation 1: $5x - 2y - 4z = 3$

Equation 2: $6x + 6y + 4z = -6$

22.3 Add the resulting equations to eliminate z,

Add the resulting equations to eliminate z,

and add two remaining equations to produce "Equation 4".

Equation 4: $11x + 4y = -3$

22.4 Reduce the system to 2 equations and 2 variables

22.4.1 Eliminate variable z in a second set of equations.

Reduce the system to 2 equations and 2 variables

This time use Equation 2 and Equation 3.

Follow the same process as above. We eliminate the same variable z .

Choose Equation 2 and Equation 3

Equation 2: $3x + 3y + 2z = -3$ Multiply by -3

Equation 3: $-2x + 5y + 3z = 3$ Multiply by 2

Multiply thru to get new equivalent equations

Equation 2b: $-9x + -9y - 6z = 9$

Equation 3b: $-4x + 10y + 6z = 6$

Add Equation 2b and 3b to produce Equation 5

Equation 5: $-13x + y = 15$

22.5 Add equations 4 and equation 5 to eliminate y, and solve for x.

Add equations 4 and equation 5 to eliminate y

Equation 4: $11x + 4y = -3$ No change

Equation 5: $-13x + y = 15$ Multiply by -4

Equation 4b: $11x + 4y = -3$

Equation 5b: $52x - 4y = -60$

22.6 Result of adding equation 4b and equation 5b

63x=-63

x=-1

22.7 Back substitute value of x in equation 5

Equation 5: $-13x + y = 15$

Equation 5: $-13(-1) + y = 15$

Equation 5: $13 + y = 15$

$y = 2$

22.8 Back substitute the values of x and y in equation 2 to solve for z

Equation 2: $3x + 3y + 2z = -3$

Equation 2: $3(-1) + 3(2) + 2z = -3$

Equation 2: $3 + 2z = -3$

2z=-6

z=-2

22.9 Back substitute the values of x,y and z in equation 2 to check the solution

22.9.1 Equation 1 or 2 or 3 can be used to verify the solution

The solution is (x, y, z)=(-1, 2, -3)

Test answer using equation 1

Equation 1: $5x - 2y - 4z = 3$

5(-1)-2(2)-3(-3)=3

-9+12=3

3=3 It checks out.

Chapter 23

Cramer's rule set up

23.1 Cramer's rule is a method to solve a linear equations system

Cramer's rule is a method to solve a linear equations system by using the coefficients of the x, y, c in a two-equation system and x, y, z, c in a three-equation system

Definition: An augmented matrix is a matrix that has the x, y, and constant in an ordered matrix.

Definition-2: A determinate D uses the x and y coefficients

Definition-3: A determinate Dx used the y and constant values

Definition-4: A determinate Dy used the x and constant values

Where the constant value is on the right side of the equals sign

Note: The D uses the values on the left side of the equation.

Note-2: The Dx uses the remainders of the columns not x.

Note-3: The Dy uses the remainders of the columns not y.

$$D = \begin{pmatrix} x_1 & y_1 \\ x_2 & y_2 \end{pmatrix}$$

$$D_x = \begin{pmatrix} c_1 & y_1 \\ c_2 & y_2 \end{pmatrix}$$

$$D_y = \begin{pmatrix} x_1 & c_1 \\ x_2 & c_2 \end{pmatrix}$$

The determinate of a 2x2 matrix is

$$\begin{bmatrix} a_1 & b_1 \\ a_2 & b_2 \end{bmatrix}$$

And is defined as

$$D = \begin{bmatrix} a_1 & b_1 \\ a_2 & b_2 \end{bmatrix}$$

Its value is the difference cross product of its values starting with

$$a_1 b_2 - a_2 b_1$$

Therefore:

$$D = \begin{bmatrix} a_1 & b_1 \\ a_2 & b_2 \end{bmatrix} = (a_1 b_2 - a_2 b_1)$$

The augmented matrix is used to solve two linear equations in a system of two equations.

$$x = \frac{D_x}{D} \; and \; y = \frac{D_y}{D}$$

The second solution involves using Cramer's rule.
The determinate of a 2 x 2 is:

$$\begin{bmatrix} a_1 & b_1 \\ a_2 & b_2 \end{bmatrix} \quad \text{and is defined by} \quad D = \begin{bmatrix} a_1 & b_1 \\ a_2 & b_2 \end{bmatrix}$$

and the values needed are:

$$D = \begin{bmatrix} a_1 & b_1 \\ a_2 & b_2 \end{bmatrix}$$

$$D_x = \begin{bmatrix} c_1 & y_1 \\ c_2 & y_2 \end{bmatrix}$$

$$D_y = \begin{bmatrix} x_1 & c_1 \\ x_2 & c_2 \end{bmatrix}$$

Chapter 24

Cramer's Rule Calculations

Solve the system of two equations by Cramer's Rule

$$3x + 11y = -9$$
$$7x + 6y = 38$$

$$\begin{vmatrix} x_1 + y_1 = c_1 \\ x_2 + y_2 = c_2 \end{vmatrix}$$

Use the above format

$$\begin{vmatrix} 3x + 11y = -9 \\ 7x + 6y = 38 \end{vmatrix}$$

$$x = \frac{D_x}{D} = \frac{\begin{bmatrix} c_1 & b_1 \\ c_2 & b_2 \end{bmatrix}}{\begin{bmatrix} a_1 & b_1 \\ a_2 & b_2 \end{bmatrix}} = \frac{(c_1 b_2 - c_2 b_1)}{(a_1 b_2 - a_2 b_1)}$$

$$y = \frac{D_y}{D} = \frac{\begin{bmatrix} a_1 & c_1 \\ a_2 & c_2 \end{bmatrix}}{\begin{bmatrix} a_1 & b_1 \\ a_2 & b_2 \end{bmatrix}} = \frac{(a_1 c_2 - a_2 c_1)}{(a_1 b_2 - a_2 b_1)}$$

$$x = \frac{D_x}{D} = \frac{\begin{bmatrix} c_1 & b_1 \\ c_2 & b_2 \end{bmatrix}}{\begin{bmatrix} a_1 & b_1 \\ a_2 & b_2 \end{bmatrix}} = \frac{(c_1 b_2 - c_2 b_1)}{(a_1 b_2 - a_2 b_1)} = \frac{((-9 * 6) - (38 * 11))}{((3 * 6) - (7 * 11))} = \frac{-472}{-59} = 8$$

$$y = \frac{D_y}{D} = \frac{\begin{bmatrix} a_1 & c_1 \\ a_2 & c_2 \end{bmatrix}}{\begin{bmatrix} a_1 & b_1 \\ a_2 & b_2 \end{bmatrix}} = \frac{(a_1c_2 - a_2c_1)}{(a_1b_2 - a_2b_1)} = \frac{((3*38)-(7*-9))}{((3*6)-(7*11))} = \frac{177}{-59} = -3$$

Therefore, the solution is x, y = (8, -3)

24.1 Easier solution by calculating individually

$$D = \begin{pmatrix} x_1 & y_1 \\ x_2 & y_2 \end{pmatrix} = -59$$

$$D_x = \begin{pmatrix} c_1 & y_1 \\ c_2 & y_2 \end{pmatrix} = -472$$

$$D_y = \begin{pmatrix} x_1 & c_1 \\ x_2 & c_2 \end{pmatrix} = 177$$

$$x = \frac{D_x}{D} = \frac{\begin{bmatrix} -9 & 11 \\ 38 & 6 \end{bmatrix}}{\begin{bmatrix} 3 & 11 \\ 7 & 6 \end{bmatrix}} = \frac{(-54-418)}{(18-77)} = \frac{-472}{-59} = +8$$

$$y = \frac{D_y}{D} = \frac{\begin{bmatrix} 3 & -9 \\ 7 & 38 \end{bmatrix}}{\begin{bmatrix} 3 & 11 \\ 7 & 6 \end{bmatrix}} = \frac{(114-(-63))}{(18-77)} = \frac{177}{-59} = -3$$

24.2 Linear system 2 equations 8,-3

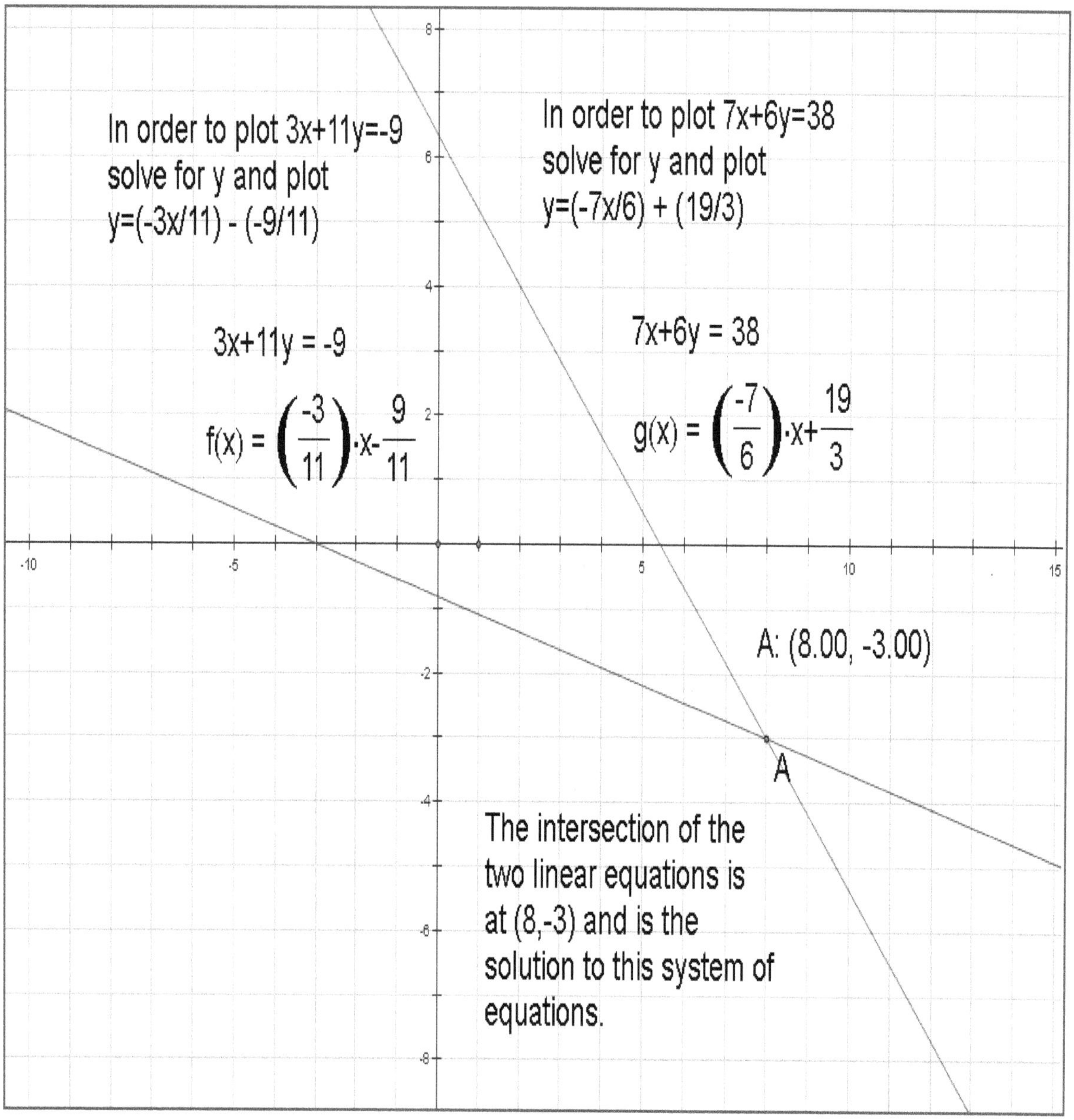

Chapter 25

3 equations and 3 unknowns solved by Cramer's rule

25.1 3x3 General Formula

This is the same problem solved by simultaneous equations earlier on page 79 using a different method. Successive addition of 3 equation system.

General Format

$$\begin{array}{ll} Equation1 & Ax_1 + By_1 + Cz_1 = D_1 \\ Equation2 & Ax_2 + By_2 + Cz_2 = D_2 \\ Equation3 & Ax_3 + By_3 + Cz_3 = D_3 \end{array}$$

System to be solved

$$\begin{array}{ll} Equation1 & 5x - 2y - 4z = 3 \\ Equation2 & 3x + 3y + 2z = -3 \\ Equation3 & -2x + 3y + 2z = -3 \end{array}$$

25.2 3x3 system solution using determinates

D is the determinate for the coefficients on the left side of the equation set.

D_x Replace the x column with the right side of the equation system.

D_y Replace the y column with the right side of the equation system.

D_z Replace the z column with the right side of the equation system.

25.3 solve for x, y, z, using determinates

$$x = \frac{D_x}{D}$$

$$y = \frac{D_y}{D}$$

$$z = \frac{D_z}{D}$$

25.4 In general, a 2 by 2 matrix is solved using this formula

In general, a 2 by 2 matrix is solved using this formula

$$D_{2x2} = \begin{bmatrix} a_1 & b_1 \\ a_2 & b_2 \end{bmatrix} = (a_1 b_2 - a_2 b_1)$$

25.5 In general, a 3 by 3 matrix is solved using this formula

In general, a 3 by 3 matrix is solved using this formula.
Notice that the operators independent
of the values used are $+, -, +$
In other words, the second operator is a minus [-] independent of the of the values used in a particular problem.

$$D_{3x3} = \begin{vmatrix} a_1 & b_1 & c_1 \\ a_2 & b_2 & c_2 \\ a_3 & b_3 & c_3 \end{vmatrix} = a_1 \begin{bmatrix} b_2 & c_2 \\ b_3 & c_3 \end{bmatrix} - a_2 \begin{bmatrix} b_1 & c_1 \\ b_3 & c_3 \end{bmatrix} + a_3 \begin{bmatrix} b_1 & c_1 \\ b_2 & c_2 \end{bmatrix}$$

25.6 Specific 3 x 3 determinate for Cramer's solution.

Specific 3 x 3 determinate for Cramer's solution.

$$D = \begin{vmatrix} a_1 & b_1 & c_1 \\ a_2 & b_2 & c_2 \\ a_3 & b_3 & c_3 \end{vmatrix} = a_1 \begin{bmatrix} b_2 & c_2 \\ b_3 & c_3 \end{bmatrix} - a_2 \begin{bmatrix} b_1 & c_1 \\ b_3 & c_3 \end{bmatrix} + a_3 \begin{bmatrix} b_1 & c_1 \\ b_2 & c_2 \end{bmatrix}$$

Equation 1 $5x{-}2y - 4z = 3$
Equation 2 $3x + 3y + 2z = -3$
Equation 3 $3x + 3y + 2z = -3$

Column | x y z| c |

$$D = \begin{vmatrix} 5 & -2 & -4 & | & 3 \\ 3 & 3 & 2 & | & -3 \\ -2 & 5 & 3 & | & 3 \end{vmatrix}$$

25.7 Resolved matrix using the coefficients of the constant in the x, y, z, columns for D, D_x, D_y, D_z

Calculate D

$$D = \begin{vmatrix} 5 & -2 & -4 \\ 3 & 3 & 2 \\ -2 & 5 & 3 \end{vmatrix} =$$

$$D = 5 \begin{bmatrix} 3 & 2 \\ 5 & 3 \end{bmatrix} - \begin{bmatrix} -2 & -4 \\ 5 & 3 \end{bmatrix} + (-2) \begin{bmatrix} -2 & -4 \\ 3 & 2 \end{bmatrix} =$$

$3[(3 \bullet 3) - (5 \bullet 2)] + 3[(-2 \bullet 3) -$

$(5 \bullet (-4))] + (-2)\,[((-2) \bullet 2) - (3 \bullet 4)]$

$5[(9 - 10] \quad - 3[-6 + 20] \quad + (-2)\,[(-4) - (-12)]$

$(-5) \quad + (-42) \quad + (-16) = -63 = D$

$D = -63$

Calculate D_x

$$D_x = \begin{vmatrix} 3 & -2 & -4 \\ -3 & 3 & 2 \\ 3 & 5 & 3 \end{vmatrix}$$

$$D_x = 3\begin{bmatrix} 3 & 2 \\ 5 & 3 \end{bmatrix} - (-3)\begin{bmatrix} -2 & -4 \\ 5 & 3 \end{bmatrix} + 3\begin{bmatrix} -2 & -4 \\ 3 & 2 \end{bmatrix}$$

$3[(3 \bullet 3) - (5 \bullet 2)] \ + 3[(-2 \bullet 3) -$

$(5 \bullet (-4))] \ + (3)\,[((-2) \bullet 2) - (3 \bullet 4)]$

$3(-1) + 3(14) + 3\,(8) = -3 + 42 + 24 = 63 = D_x$

$D_x = 63$

Calculate D_y

$$D_y = \begin{vmatrix} 5 & 3 & -4 \\ 3 & -3 & 2 \\ -2 & 3 & 3 \end{vmatrix}$$

$$D_y = 5\begin{bmatrix} -3 & 2 \\ 3 & 3 \end{bmatrix} - 3\begin{bmatrix} 3 & -4 \\ 3 & 3 \end{bmatrix} + (-2)\begin{bmatrix} 3 & -4 \\ -3 & 2 \end{bmatrix}$$

$5[(-3 \bullet 3) - (3 \bullet 2)] \ + 3[(3 \bullet 3) - (3 \bullet (-4))] \ + (-2)\,[(3 \bullet 2) - (-3 \bullet -4)]$

$5\,(-9 - 6) \quad - 3\,(9 - (-12)) \quad + (-2)\,(6 - 12)$

$-75 - 63 + 12 = -126 = D_y$

$D_y = -126$

Calculate D_z

$$D_z = \begin{vmatrix} 5 & -2 & 3 \\ 3 & 3 & -3 \\ -2 & 5 & 3 \end{vmatrix}$$

$5[(3 \bullet 3) - (5\,(-3))] \ - 3[(-2 \bullet 3) - (5 \bullet 3)] \ + (-2)\,[(-2 \bullet -3) - (3 \bullet 3)]$

$5[24] - 3[-21] + (-2)\,[-3]$

$120 + 63 + 6 = 189 = D_z$

$D_z = 189$

25.8 Solution to determinate problem

$$x = \frac{D_x}{D}$$

$$y = \frac{D_y}{D}$$

$$z = \frac{D_z}{D}$$

$$x = \frac{D_x}{D} = \frac{63}{-63} = -1$$

$$y = \frac{D_y}{D} = \frac{-126}{-63} = 2$$

$$z = \frac{D_z}{D} = \frac{189}{-63} = -3$$

The solution is (x, y, z,)=(-1,2,-3)

Test answer using equation 1

Equation 1:

$$5x - 2y - 4z = 3$$

5(-1)-2(2)-3(-3)=3

-9+12=3 —item 3=3 It checks out.

Chapter 26

Inequalities and Absolute value

$|x| = c$ Absolute value Equals c
$|x| \geq c$ Absolute value Greater than or equal to c
$|x| > c$ Absolute value Greater than c
$|x| \leq c$ Absolute value Less than or equal to c
$|x| < c$ Absolute value Less than c

The arrow points to the smaller value.

26.1 Absolute Values shown as Inequalities and Equals on Real Number Line.

- Inequalities are Ranges whereas Equals are Points on a real number line
- Define x, absolute x and the resultant c and graph on a real number line.
- Absolute values are shown bounded by two vertical lines $|x|$ with the result in this general case represented by c
- Since x is within the absolute value brackets $|x|$, x can be either positive or negative, and c is always positive. C represents any real number.
- The result of an absolute value is always positive therefore: $|-x| = c$ and $|x| = c$

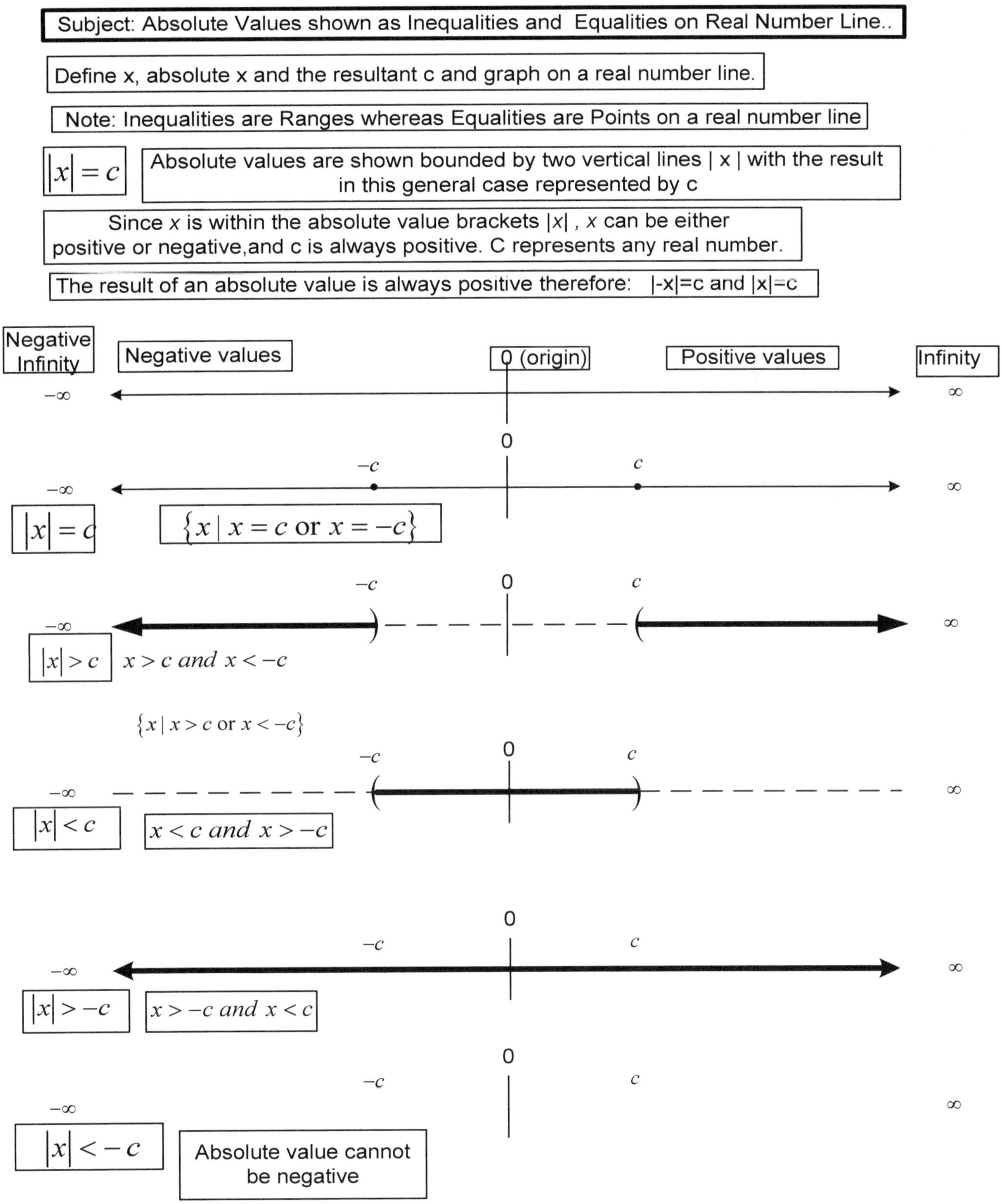

Figure 26.1: Absolute-Value-inequality-chart

Chapter 27

Absolute Value and Inequality calculations

Refer to graph on page 101 to show this example.

- Absolute-Value-inequality calculations for $|3 - 2x| > 7$
- Remove the absolute value bars |(function)|
- $|3 - 2x| > 7$ Means $3 - 2x > 7 \quad \text{and} \quad 3 - 2x < -7$
- There are two solutions to an absolute value expression
- Solution 1 : $3 - 2x > 7$ No Change
- Solution 2 : $3 - 2x > 7$ Multiply by -1 to get absolute value
- $(-1)\,(3 - 2x > 7) = \quad -(3 - 2x) < (-7) =$
- Solution 2: $-3 + 2x < -7$
- Solution 1: $3 - 2x = 7; x = -2$; Left intersecting point
- Solution 2: $-(3 - 2x) = 7; x = 5$; Right intersecting point
- Note the change of direction of the in-equals sign
- Multiplying (or dividing) by -1 (minus 1) changes the direction if the inequality sense.
- In this case multiplying by -1 changed the sign from "greater than" " $>$ " to "less than" $<$ ".
- Given two solutions both solutions are now resolved.

27.1 The solutions represent the projection on the real number line.

- The solutions represent the projection on the real number line, of intersection of the absolute value equation , and the "y" value it tested against.
- Solve for the projection of the intersection of y=7 and

  $$|3 - 2x|$$

  by setting each side of the equation to 7 and solve for x.
- Solution 1: $3 - 2x = 7; x = -2$; Left intersecting point
- Solution 2: $-(3 - 2x) = 7; x = 5$; Right intersecting point

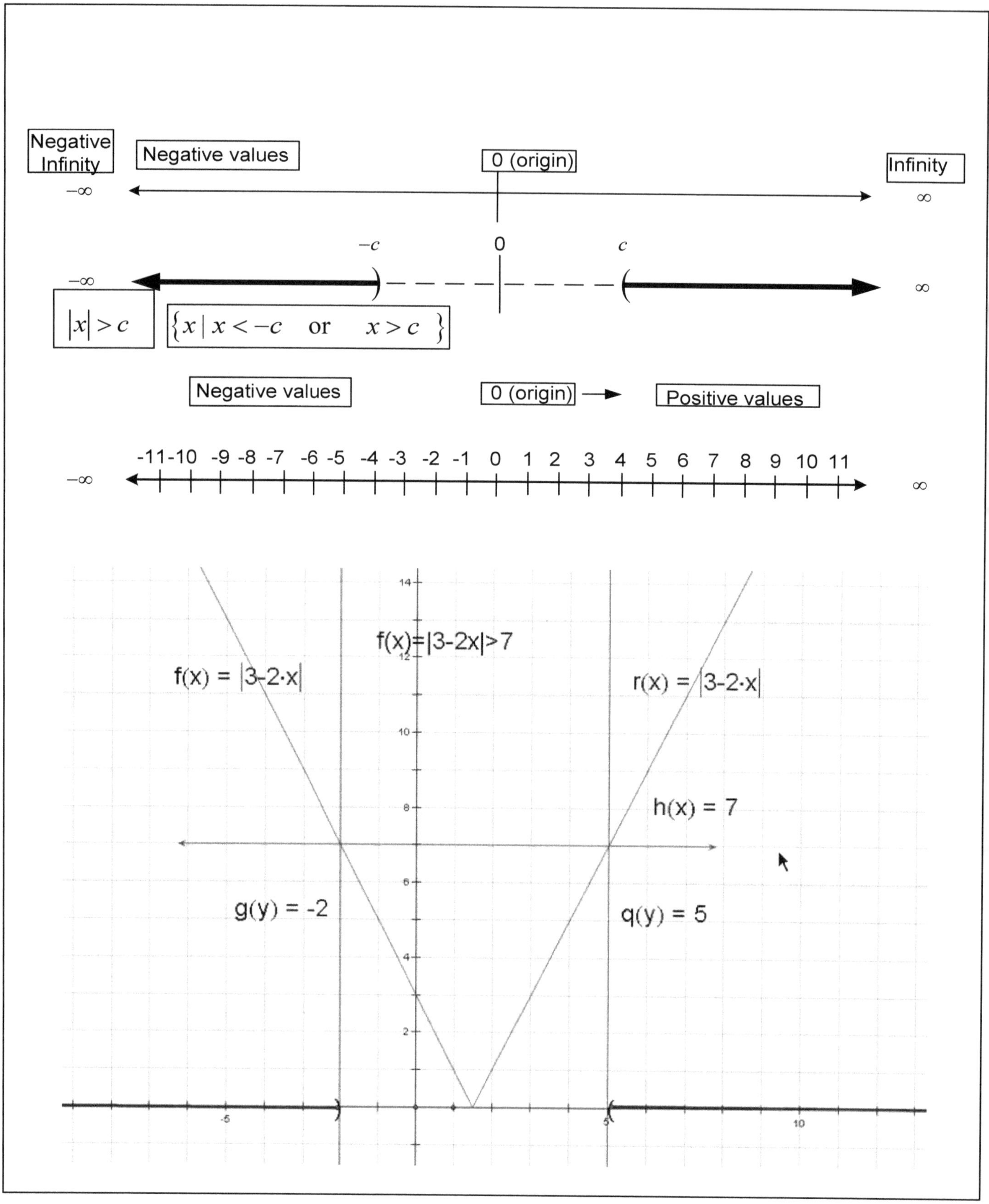

Figure 27.1: Absolute-Value-inequality

Chapter 28

Problem Solving Six Step Process

28.1 Applied Algebra

1. Step 1 **Read the problem carefully**
   until you understand what is given and what is to be found.
2. Step 2 **Assign a variable** to represent the unknown value.
   (a) Use diagrams or tables as needed.
   (b) Write down what the variable represents.
   (c) Then, express any other unknown values in terms of the first variable.
3. Step 3 **Write an equation** using the variable expression(s).
4. Step 4 **Solve the equation.**
5. Step 5 **State the answer.**
6. Step 6 **Check the answer** in the words of the original problem.
   (a) Does the solution seem reasonable?
7. Step 6 **Check the answer** in the words of the original problem.

28.2 Garden problem, Application of Six step rule process

Steps $1-6$ **Math Solution as a word problem.**

1. **Step 1 Read the problem carefully**

   until you understand what is given and what is to be found.

   The width of the garden will be 4 feet less than the garden's length.

   We need to find the dimensions of a garden with area 96 ft^2**.**

2. **Step 2 Assign a variable to represent**

   the unknown value, using diagrams or

   tables as needed. Write down what the variable represents.

   Then, express any other unknown values in terms of the first variable.

   Let x = the length of the garden. Then x - 4 = the width.

   (The width is 4 ft less than the length.)

3. **Step 3 Write an equation using the variable expression(s).**

   Area=Length times Width $A = LW$**therefore** $96 = x(x-4)$

4. **Step 4 Solve the equation. Formula** $96 = x(x-4)$

   Multiply and solve for Zero $0 = x^2 - 4x - 96$

   Factor $(x-12)(x+8) = 0$

   Set each factor equal to zero and solve $(x-12) = 0$ **Then** $x = 12$ **Set remaining factor equal to zero and solve** $(x+8) = 0$

   Then $x = -8$

5. **Step 5 State the answer.**

   Does it seem reasonable?

Given that the answers are -8 and 12 we discard the -8 since dimensions are positive.

We accept the 12 answer since it is positive.

6. **Step 6**

   Check the answer in the words of the original problem.

   If $x = 12$ **then** $(x - 4) = (12 - 4) = 8$

   Then the dimensions are

   Width=8 and Length =12.

   If A=lw then 96=(8)(12)=96 so the answer checks out.

Steps $1 - 6$

Math Solution in an Algebra formulas progression.

Solution

$$A = LW$$

$$96 = x(x - 4)$$

$$0 = x^2 - 4x - 96$$

$$(x - 12)(x + 8) = 0$$

$$(x - 12) = 0$$

$$x = 12$$

$$(x + 8) = 0$$

Not used x=-8

$$x = -8$$

Accepted x=12

$$x = 12$$

Check:

$$(x - 4) = (12 - 4) = 8$$

$$A - LW$$

$$96 = (8)(12) = 96$$

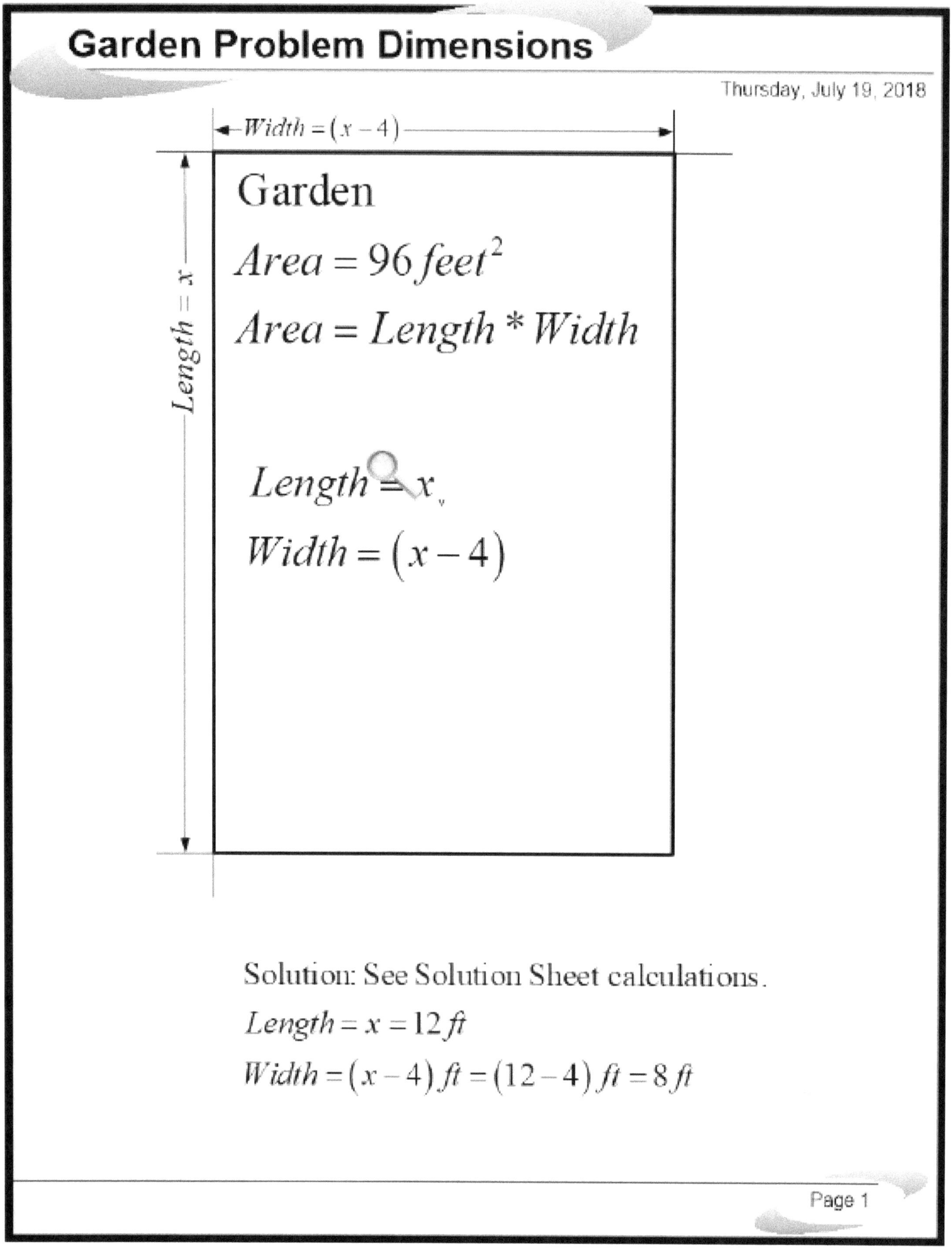

Figure 28.1: Garden Problem Dimensions

Chapter 29

More Problem Solving in Algebra-1

29.1 Calculate two-person time to complete 1 spraying job.

Sam and Chester can complete a spraying job together but at different rates

1. Sam takes 8 hours to do the work alone
2. Sam's Rate per hour is 1/8 of the job in one hour
3. Chester takes 14 hours to do the work alone
4. Chester's Rate per hour is 1/14 of the job in one hour
5. The time working together is x the same for both Sam and Chester

Worker	Rate	Working Together Time	Factional Part of Work Done
Sam	1/8	x	1/8 x
Chester	1/14	x	1/14 x

1. Rate per hour equals work done in one hour.
2. Time equals x
3. Work equals rate * time
4. Work in entire job equals 1

5. Factional part Sam plus fractional part Chester equals 1 job

$$\frac{1}{8}x + \frac{1}{14}x = 1 \quad \text{The LCD} = (8)(14) = 56$$

$$(56) * \frac{1}{8}x + (56) * \frac{1}{14}x = (56) * 1$$

$$7x + 4x = 56; \quad 11x = 56$$

$x = \dfrac{56}{11} = 5\dfrac{1}{11}$ hours to complete the job of spraying. Answer

Chapter 30

Two Cars algebra word problem

30.1 Solving word problems using equations

30.2 Two car problem in words

Two cars leave Baton Rouge LA at the same time and travel North. Car A travels at 63 mph and Car B travels at 55 mph. In how many hours will be 24 miles apart?

30.3 Analyze the problem

These problems are real life problems solved by the Algebra process.
The first thing to notice is that the time is the same in the problem.
The second is that distance = rate x time.
Also, since this problem cannot easily be done in one step, we need to set up a 2 x 3 matrix of everything we know.
Distance in this case will not initially be solve initially, but we can express distance in Algebraic terms. i.e. 63t and 55t.

	Rate *	Time	= Distance
Car A	63 mph	T	63T
Car B	55 mph	T	55T

Note mph = miles per hour or $\frac{\text{miles}}{\text{hour}}$

The units in this equation:

Rate: mph in miles

Time: time in hours

Distance = Time in miles; D=rt or rt=d

Miles per hour $=\frac{\text{Miles}}{\text{Hour}}$ or mph

mph * Hours = Miles

$$\frac{\text{Miles}}{\text{Hour}} * \text{Hours} = \text{Miles}$$

30.4 Formula and Solution

Describe the problem in words

Distance Car A minus (-) Distance Car B= 24 Miles

Translate the words description into an algebraic equation

Use the formula distance = rate*time d=rt

63T-55T=24 Miles

Solve for T hours

8T=24

T=3 hours

Check :
$63 \bullet 3 = 189$ miles
$55 \bullet 3 = 165$ miles
189 miles - 165 miles = 24 miles
It checks

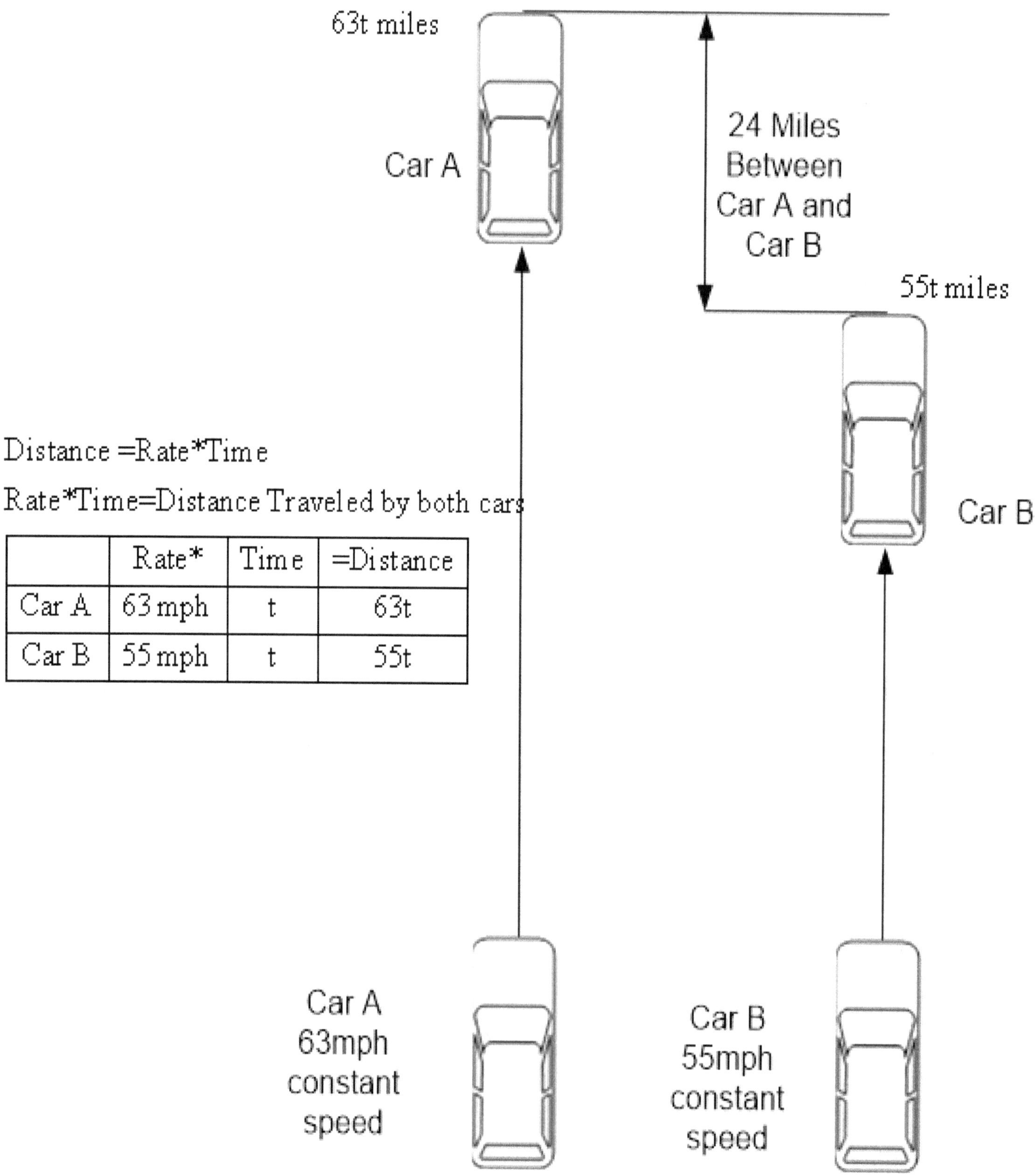

	Rate*	Time	=Distance
Car A	63 mph	t	63t
Car B	55 mph	t	55t

Figure 30.1: Graphical Solution for two cars

Chapter 31

Thrown ball from a roof top: Quadratic Equation

31.1 Calculate the time it takes to hit the ground

A ball is thrown from a roof top with an initial speed of 16 feet/second from the rooftop at 192 feet.

The following equation derivation (from physics) is found on on page 167

$$s(t) = -16t^2 + 16t + 192$$
$$s(t) \text{ is the height dependent on the time } (t)$$

$$\text{At ground level } s(t) = 0$$

$$0 = -16t^2 + 16t + 192$$

solve by dividing both sided by -4

$$4t^2 - 4t - 48 = 0$$

$$(4t + 12)(t - 4) = 0$$

$$4t + 12 = 0$$
$$4t = -12$$
$$t = -4 \text{ reject a negative time solution}$$

$$(t - 4) = 0$$
$$t = 4 \quad \text{seconds for the ball to hit the ground}$$

31.2 Calculate the height of the vertex

$$x_{vertex} = \left(\frac{-b}{2a}\right)$$

$$\left(\frac{-b}{2a}\right)$$

$$a = -16, b = 16$$

$$x_{vertex} = \left(\frac{-b}{2a}\right) = \frac{(-16)}{2\,(16)} = \frac{1}{2}$$

$$s\,(t) = -16t^2 + 16t + 192$$

$$s\,(t) = y = -16\left(\frac{1}{2}\right)^2 + 16\left(\frac{1}{2}\right) + 192$$

$$y = -4 + 8 + 192 = 196 \text{ feet above ground}$$

$$\text{The coordinates set of the apex is } \left(\frac{1}{2}, 196\right)$$

31.3 Calculate the time it takes to reach 96 feet

The following equation derivation (from physics) is found on page 167.

A ball is thrown from a roof top with an initial speed of 16 feet/second from the rooftop at 192 feet.

The equation from physics is:

$$s(t) = -16t^2 + 16t + 192$$

$$s(t) \;= 96 \text{ feet}$$

$$96 = \; -16t^2 + 16t + 192$$

$$0 = -16t^2 + 16t + 192 - 96$$

$$0 = -16t^2 + 16t + 96$$

divide both sides of the equation by (-4)

$$4x^2 - 4x - 24 = 0$$

$$(4x + 8)(x - 3) = 0$$

$$4x + 8 = 0$$

$$x = -2 \text{ reject solution}$$

Time is always a positive value

$$(x - 3) = 0$$

x = 3 seconds to reach 96 feet

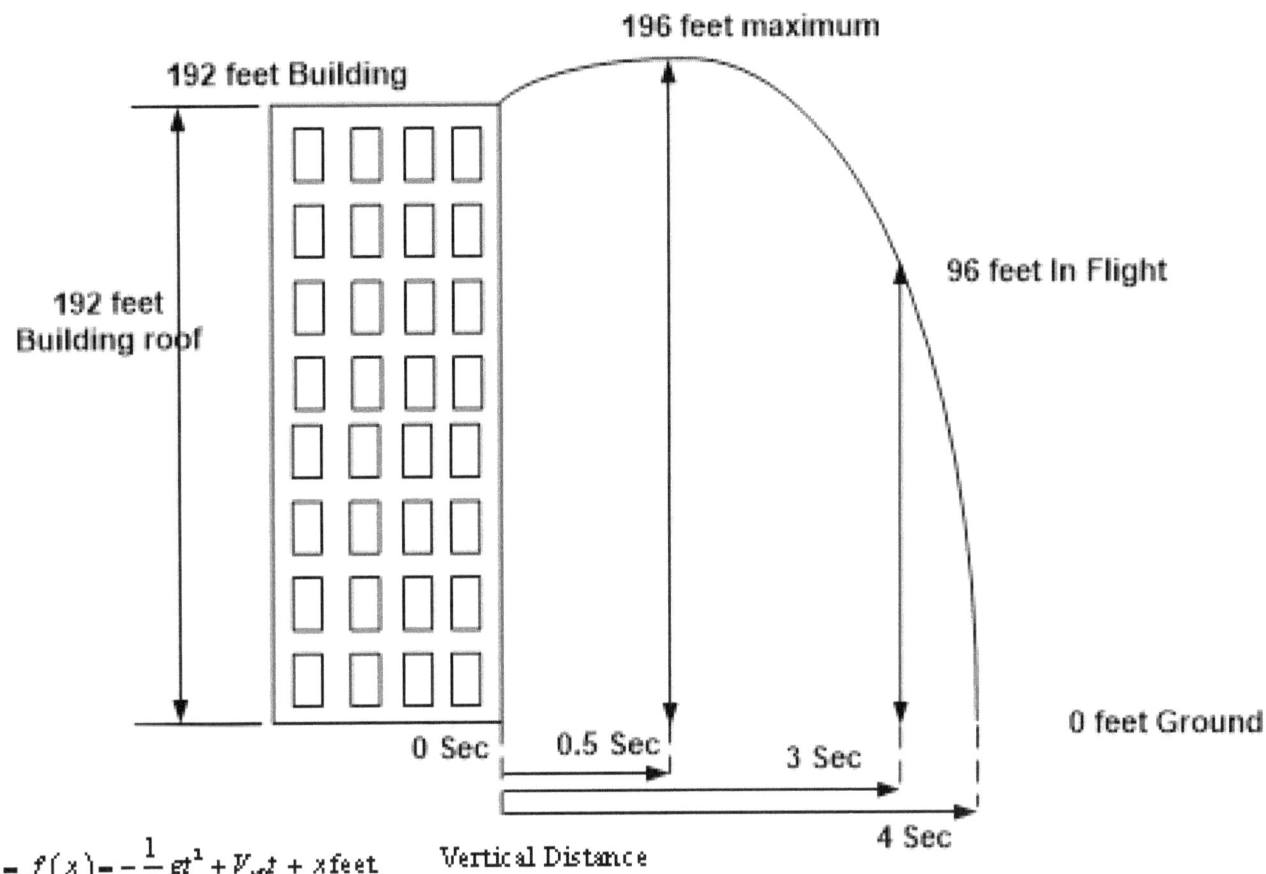

$$y = f(x) = -\frac{1}{2}gt^2 + V_{y0}t + x\text{ feet}$$

$$f(x) = -16t^2 + 16t + 192$$

$$f(x) = -16t^2 + 16t + 192$$

Time in seconds	Distance in Feet above ground
0.0	192.0
0.5	196.0
1.0	192.0
1.5	180.0
2.0	160.0
2.5	132.0
3.0	96.0
3.5	52.0
4.0	0

Figure 31.1: Vertical distance of ball thrown off roof

Chapter 32

Pythagorean-theorem

Pythagorean Theorem

In a right triangle the addition of the squares “a” and “b” equals the square of the hypotenuse “c”

$$c^2 = a^2 + b^2$$

$$5^2 = 3^2 + 4^2$$

$$25 = 9 + 16$$

$$25 = 25$$

It proves out

Note: the Pythagorean works for all right triangles.
A right triangle has 90^0 as its largest angle.
Refer to the following diagram.

A side such as a squared is represented by an area of 4 evenly sized blocks measuring 16 blocks.

A side such as b squared is represented by an area of 3 evenly sized blocks measuring 9 blocks.

A side such as c squared is represented by an area of 5 evenly sized blocks measuring 25 blocks. as shown in the diagram.

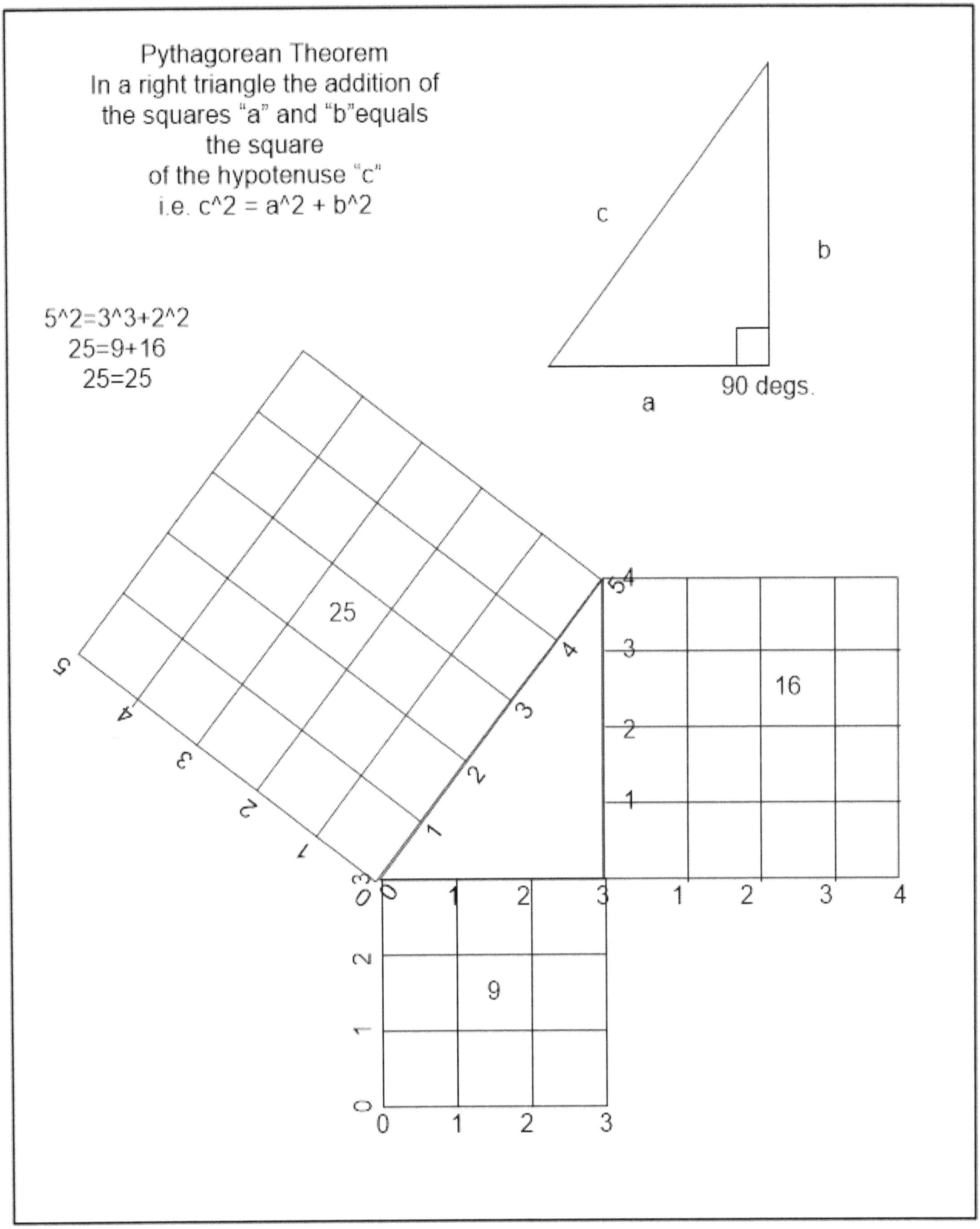

Figure 32.1: Pythagorean-theorem

Chapter 33

Ladder problem description

33.1 State the problem

- A ladder is placed on a wall.
- The ladder rests on the wall 6 feet above the floor.
- The bottom of ladder is x feet from the wall equals x
- The ladder is 2 feet longer than the distance from the wall. $x + 2$

34.1 Analyze the problem

- Let a = x equal to the short leg of the triangle. The ladder makes a right-angle triangle with the wall.
- Let b = 6 feet the height on the wall where the ladder rests on the wall.
- Let c = (x+2) the length of the ladder.

Chapter 34

Ladder problem diagram

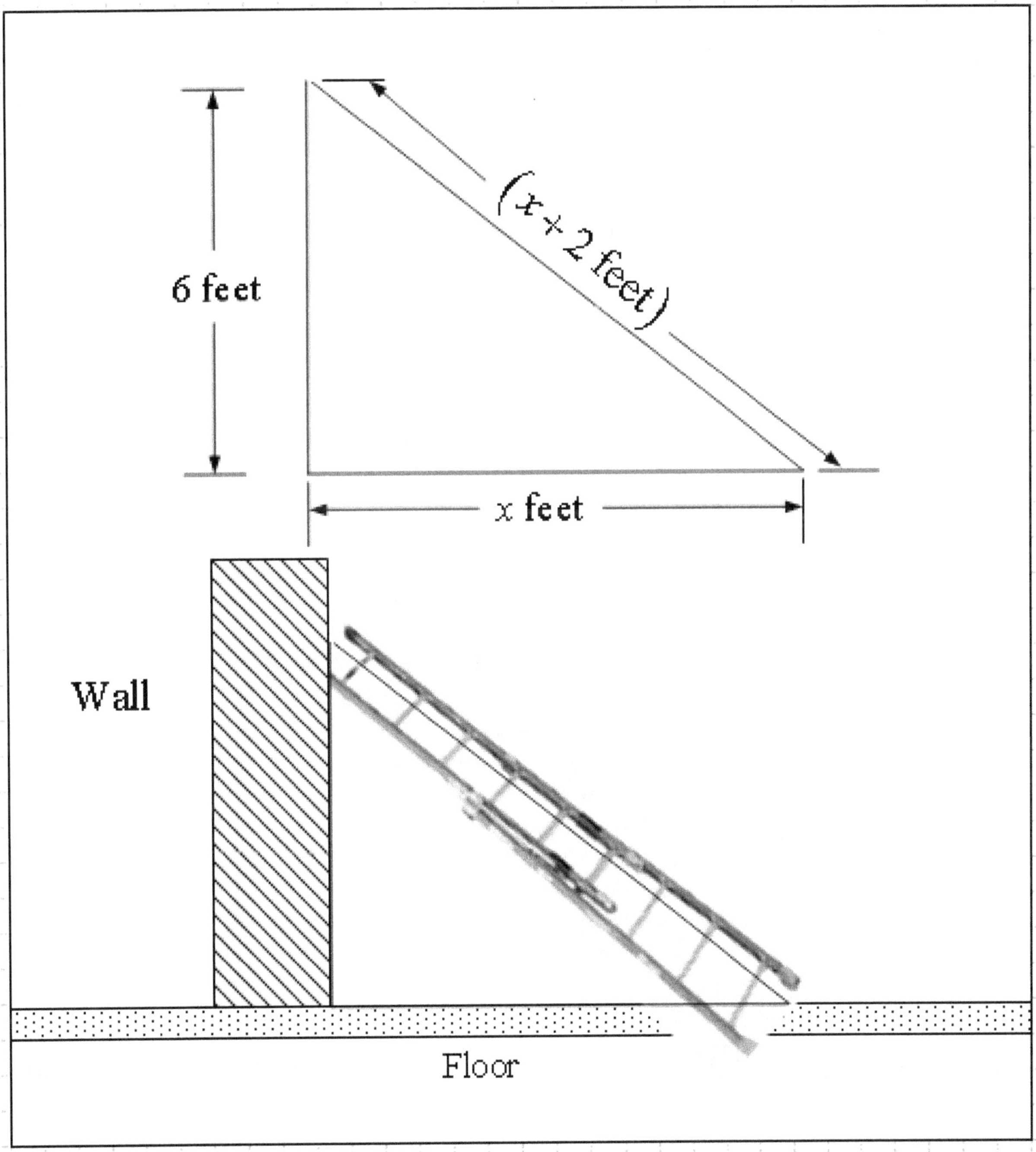

Figure 34.1: Ladder problem

Chapter 35

Ladder problem solution

- We know from the Pythagorean formula that the hypotenuse squared equals the sum of the squares of the other two sides. So we substitute the values of the triangle formed by the ladder and the wall
- This is good as far as it goes but what is asked for is the length of the ladder which equals (x+2)
- Therefore, the ladder = $(x + 2) = (8 + 2) = 10$ feet

$$x^2 + 6^2 = (x + 2)^2$$

$$x^2 + 36 = x^2 + 4x + 4$$

$$x^2 + 4x + 4 = x^2 + 36$$

$$4x + 4 = 36$$

$$4x = 32$$

$$x = 8$$

and

$$x + 2$$

$$x + 2 = 10$$

- Therefore, the ladder is 8 feet from the wall and the ladder is 10 feet long.

Chapter 36

Types of quadratic equation solutions

36.1 Example of a real solution

The quadratic formula uses the coefficients of the quadratic-equation to solve for x.

$$x = \frac{-b \pm \sqrt{b^2 - 4ac}}{2a}$$

General Format

$$ax^2 + bx + c$$

Typical Quadratic equation

$$x^2 + 4x + 4$$

Factored Expression

$$x^2 + 4x + 4 = (x+2)(x+2)$$

Where If factorable, factor into 2 binomials If not factorable use the quadratic formula Note: The Quadratic formula works in both cases.

$x^2 + 2x - 15$
$x^2 + 2x - 15 = (x + 3)(x - 5)$
Factored Form

Coefficients and constant for setting up the binomial formula
$a = 1,\ b = 2,\ c = -15$

Using the quadratic formula:

$$x = \frac{-b \pm \sqrt{b^2 - 4ac}}{2a}$$

$$x = \frac{-2 \pm \sqrt{2^2 - 4(1)(-15)}}{2(1)}$$

$$x = \frac{-2 \pm \sqrt{64}}{2}$$

$$x = \frac{-2 \pm 8}{2}; \text{ Therefore } \quad x = \frac{-2 + 8}{2} \quad x = \frac{-2 - 8}{2}$$

$$x = \frac{6}{2} = 3 \quad x = \frac{-10}{2} = -5 \quad \text{In summary } x = 3, -5$$

We can verify the solution by plotting the equation: We name the

$$\sqrt{b^2 - 4ac}$$

thc determinate of the equation.
The result of solving the determinate yields one of 3 possible solution real, imaginary or no real solution (in the case of the square root being negative.)

36.2 Imaginary number system

An imaginary number system consists of x and iy.

Where $i = \sqrt{-1}$

Imaginary numbers are plotted on the y -axis only.

A typical coordinate would be $\{x, y\} = \{5x, 3iy\}$

36.3 Example of an imaginary solution

$$3x^2 - 5x + 4 \text{ where a} = 3,\ b = -5,\ c = 4$$

$$x = \frac{-b \pm \sqrt{b^2 - 4ac}}{2a}$$

$$x = \frac{-(-5) \pm \sqrt{(-5)^2 - 4(3)(4)}}{2(3)}$$

$$x = \frac{5 \pm \sqrt{-23}}{6} = x = \frac{5 \pm 4.796\, i}{6} \quad \text{where } i \quad \sqrt{-1}$$

therefore, the roots are imaginary and do not intersect with the real number line.

36.4 Plotting a quadratic with and imaginary solution

Note: Imaginary solutions does not mean the graph of the equation does not exist. It does however mean that the parabola does not cross the real number line (the +-x-axis) hence the solution itself does not exist.

36.5 Explanation of the determinate

$$x = \frac{-b \pm \sqrt{b^2 - 4ac}}{2a}$$

The determinate is the radical part of the quadratic formula and the result determines the type of solution.
The determinate is

$$\sqrt{b^2 - 4ac}$$

- Real solutions if Determinate is greater than or equal to zero.
- If equation is equal zero then the apex (maximum or minimum) intersects with the x axis and there are 2 solutions that are the same i.e. one solution
- If the determinate is less than zero then there are imaginary solutions and the entire parabola is either above or below the x-axis.
- Note: parabolas with imaginary solutions are still able to be represented on a graph. (Cartesian coordinate graph)
- The quadratic formula as a whole solves for x depending of the coefficients of the equation.
- Calculating the expression, $\left(\frac{-b}{2a}\right)$ will solve the apex of the quadratic.
- Calculating the $\sqrt{b^2 - 4ac}$ is used to determine whether the root is real, zero, or imaginary.

36.6 Examples of quadratic equations

Examples of quadratic equations, are throwing a ball up in the air with an upward motion and calculating the results.

The problem stated, is throwing in the x-direction (upward).

An example of a vertical problem found on page 115

Chapter 37

Real Solution of a quadratic equation, solved and plotted on a cartesian coordinate graph

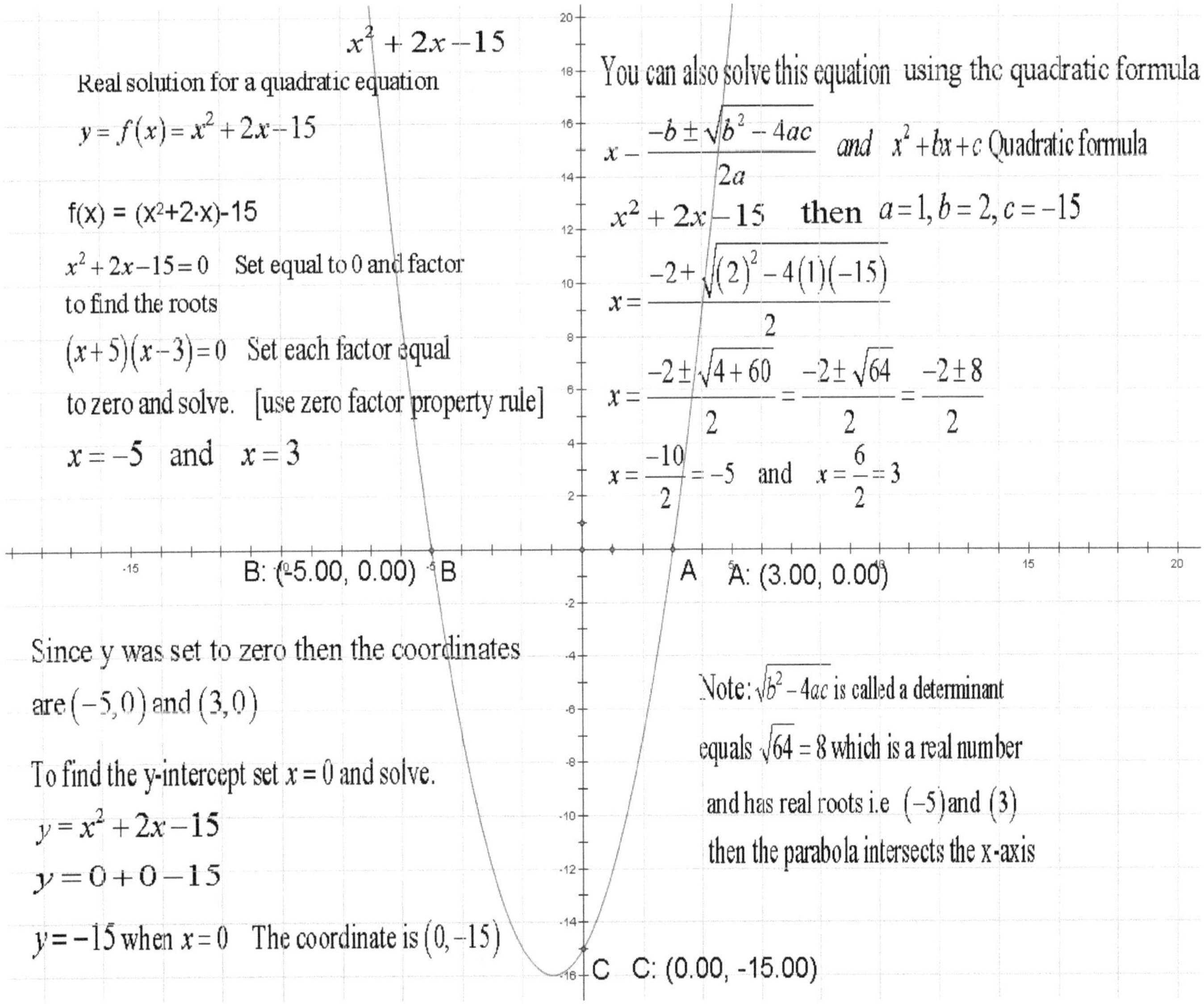

Figure 37.1: Real Solution of a quadratic equation, solved and plotted on a cartesian coordinate graph

Chapter 38

Imaginary Solution of a quadratic equation, no real intersections

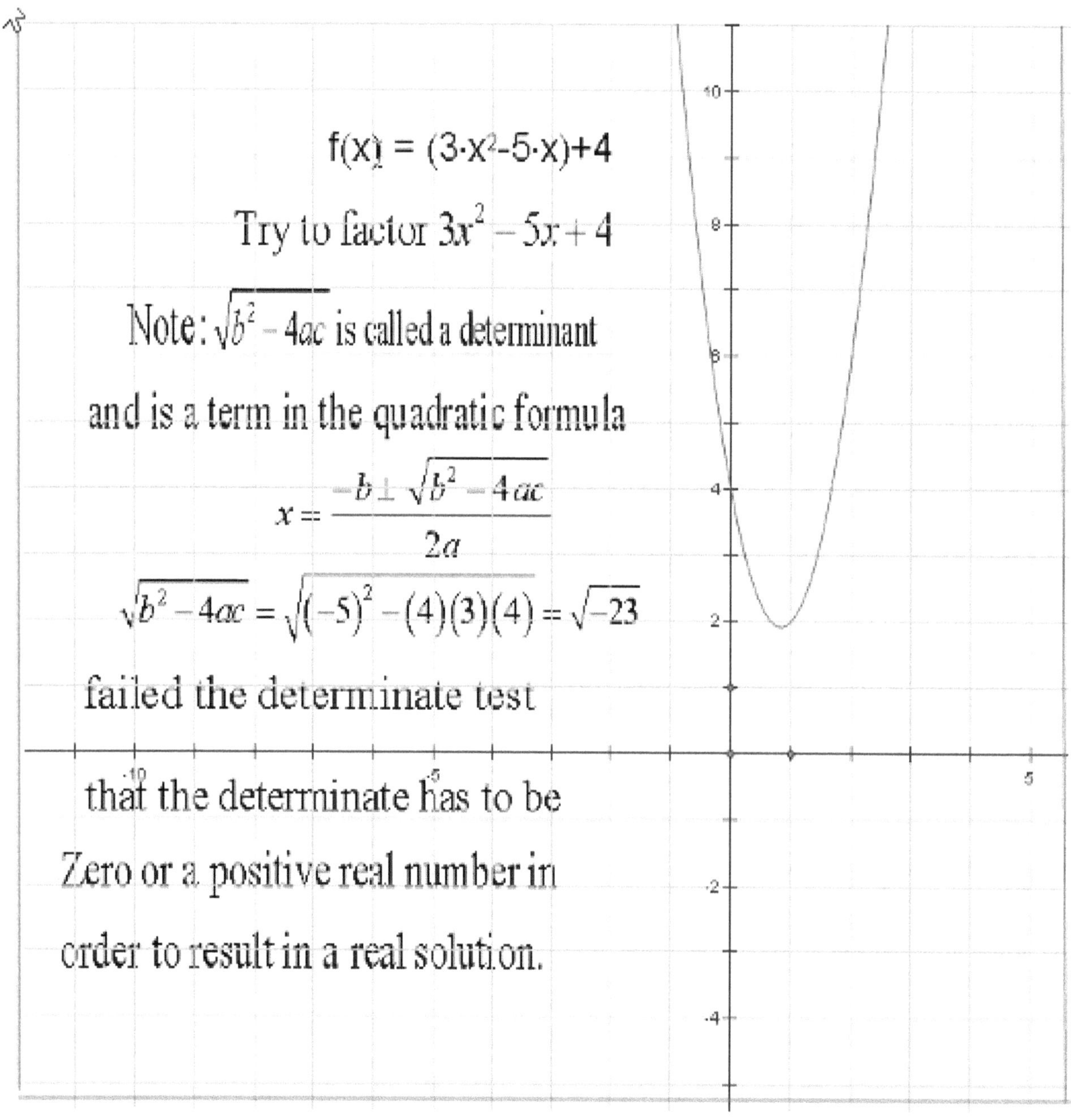

Figure 38.1: Imaginary Solution of a quadratic equation

Chapter 39

Rules of Exponents

39.1 Rules of exponents.

The m's and n's are integers exponents.
Note: The $x's$ are variables, real numbers, or algebraic expressions when working with the following rules.

Property	Examples
1) $x^m x^n = x^{m+n}$	$3^3 \bullet 3^5 = 3^{3+5} = 3^8 = 6561$
2) $\frac{x^m}{x^n} = x^{m-n}$	$\frac{x^8}{x^3} = x^{8-3} = x^5$
3) $x^{-n} = \frac{1}{x^n}$	$x^{-4} = \frac{1}{x^4}$
4) $x^0 = 1$	$\left(x^2+2\right)^0 = 1$
5) $(xy)^n = x^n y^n$	$(3x)^3 = 3^3 x^3 = 27x^3$
6) $(x^m)^n = x^{mn}$	$\left(x^4\right)^{-5} = x^{4(-5)} = x^{-20} = \frac{1}{x^{20}}$
7) $\left(\frac{x}{y}\right)^m = \frac{x^m}{y^m}$	$\left(\frac{3}{x}\right)^2 = \frac{3^2}{x^2} = \frac{9}{x^2}$
8) $\left\|x^2\right\| = \|x\|^2 = x^2$	$\left\|\left(-2^2\right)\right\| = \|-2\|^2 = (2)^2 = 4$

39.2 Rational exponents as fractional integers.

Rational exponents as fractional exponents.
If a is a real number and n is a positive integer, such that the principle nth root of a exists, then
$\mathrm{a}^{\frac{1}{n}} = \sqrt{a}$, by definition.
where $\frac{1}{n}$ is the rational exponent of a.

Also, if m is a positive integer with no common factor with n , then
$\mathrm{a}^{\frac{m}{n}} = \left(\mathrm{a}^{\frac{1}{n}}\right)^{m} = \sqrt[n]{a}^{m}$

$$\sqrt{100} = \pm 10$$
$$\sqrt{64} = \pm 8$$

Chapter 40

Square root characteristics

- The set of square coordinates are (zero, zero), or (positive, positive).
- The square root of a number is always a real number; therefore, a negative root will not yield a real number.
- Non-perfect square roots will yield irrational value numbers.
- Example of two non-perfect numbers.

$$\sqrt{2} = 1.414$$
$$\sqrt{3} = 1.732$$
$$\sqrt{20} = 4.472$$

- Note: None of the set of coordinates are in the negative x section of the graph.
- You cannot have the square root of a negative number and have a real result. (The result will be imaginary)
- Perfect positive square roots will yield Whole numbers.

40.1 Table of first five perfect square roots

x	$y=(x)=\sqrt{x}$	$(x,y) or (f(x))$
0	$y=f(0)=\sqrt{0}=0$	$(0,0)$
1	$y=f(1)=\sqrt{1}=1$	$(1,1)$
4	$y=f(4)=\sqrt{4}=2$	$(4,4)$
9	$y=f(9)=\sqrt{9}=3$	$(9,3)$
16	$y=f(16)=\sqrt{16}=4$	$(16,4)$

Figure 40.1: First five perfect square roots

40.2 Characteristics of first five perfect square roots

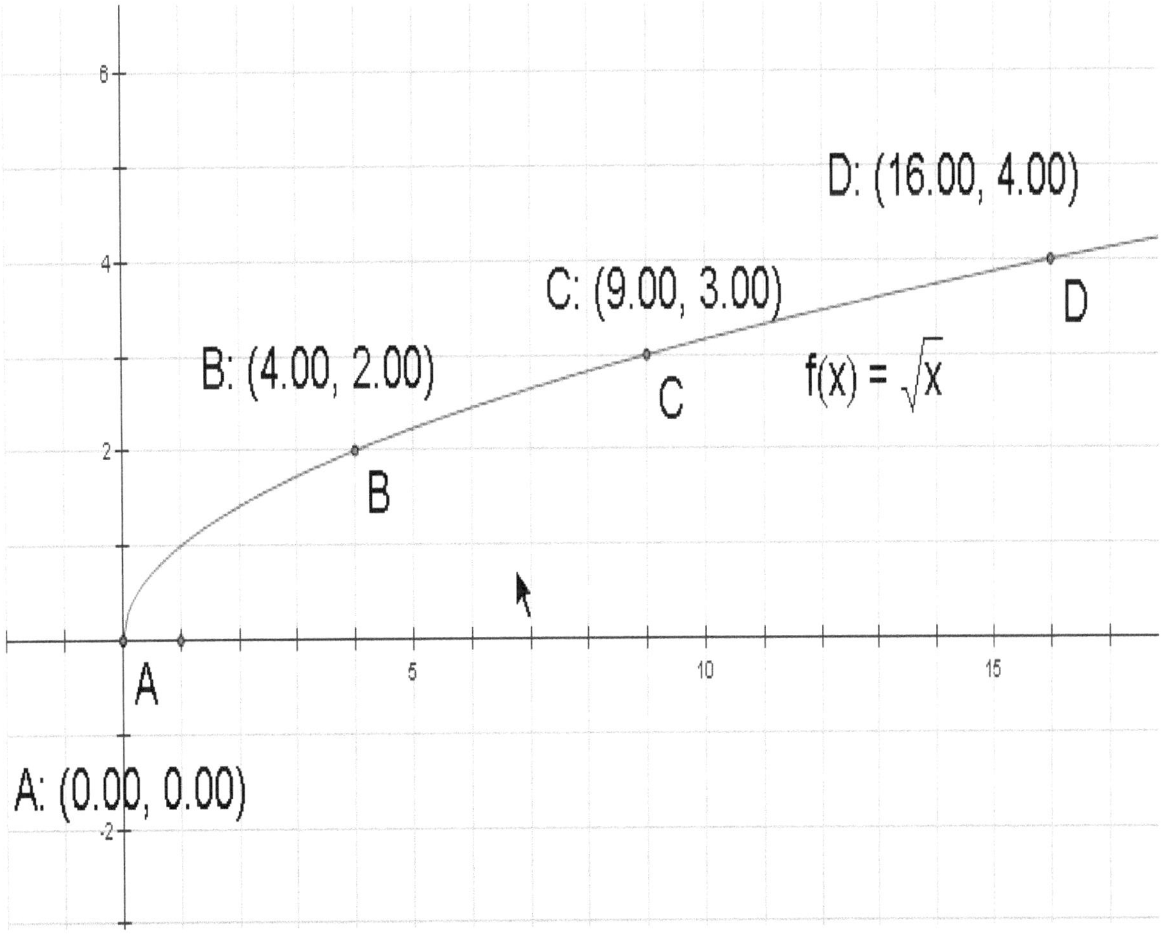

Figure 40.2: Graph of four perfect square roots.

Chapter 41

Cube root characteristics

- The cube root can be in the negative quadrant and the positive quadrant or zero.
- The set of cubic coordinates are (negative, negative), or (zero, zero), or (positive, positive).
- The cube root of a number is always a real number,therefore a negative root as well as a zero root or a positive root will yield a real number.
- Non-perfect cube roots will yield an irrational value number.
- Perfect cube roots will yield a Whole number
- The cube root of x can be expressed as a fractional exponent.
- $\sqrt[3]{x} = x^{\frac{1}{3}}$
- Therefore, we graph $x^{\frac{1}{3}}$ as a cube root. See graph.
- $h(x) \ = x^{\frac{1}{3}}$

41.1 Table of first five perfect cube roots

x	$y=(x)=\sqrt[3]{x}$	$(x,y) or (f(x))$
-8	$y=f(-8)=\sqrt[3]{-8}-2$	$(-8,-2)$
-1	$y=f(-1)=\sqrt[3]{-1}$	$(-1,-1)$
0	$y=f(0)=\sqrt{0}=0$	$(0,0)$
1	$y=f(1)=\sqrt{1}=1$	$(1,1)$
8	$y=f(8)=\sqrt[3]{8}=2$	$(8,2)$

Figure 41.1: Table of first five perfect cube roots

41.2 Graph of first five perfect cube roots

41.3 $y = \sqrt[3]{x}$ $y =$ the cube root of x

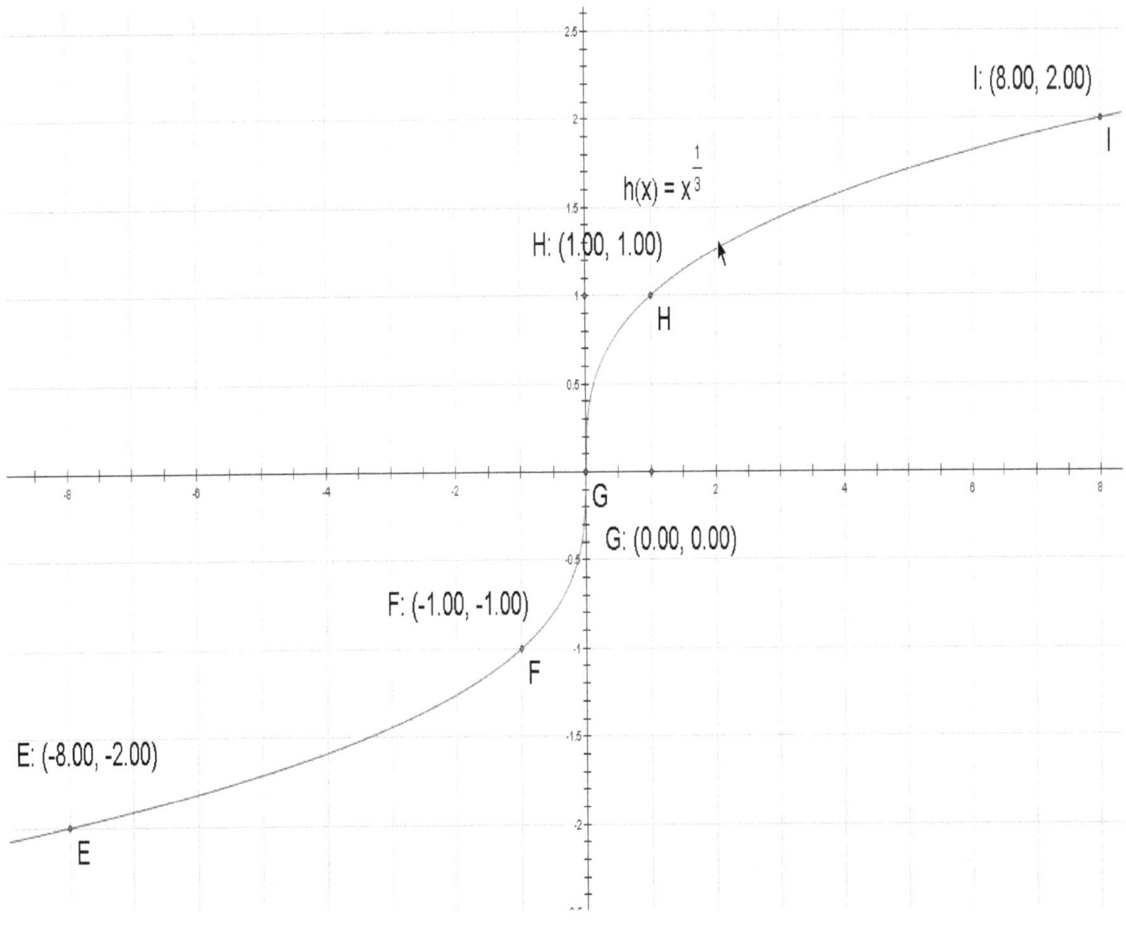

Figure 41.2: First five perfect cube roots

Chapter 42

Rational Expressions and Complex Fractions

There are two ways to solve complex fractions

- Case 1: Use Lowest common denominator separately with the numerator.

- A Case 1 Easier method of solving complex fractions is found on page 145
- Case 2: Use Lowest common denominator of the entire rational
expression..
- Case 2: follows

$$\frac{\frac{1}{m-1}+\frac{2}{m+2}}{\frac{2}{m+2}-\frac{1}{m-3}}$$

Find the lowest common denominator of all the denominators

The Denominators are (m-1) and (m-3) and (m+2),(m+2) Note" For the lowest common denominator (m+2) only gets counted once since it's highest power (exponent) is one.

- The basic principle of adding rational expressions is that rational expressions can be added or subtracted, when the denominators are the same.
- To do that we need the lowest common denominator.
- The factors are counted the least time that they appear.
- Denominators (m-1) and (m+2) and (m-3) are only counted one time.
- The factors are (m-1)(m+2)(m-3) or the lowest common denominator;
- Note" For the lowest common denominator (m+2) only gets counted once.
- If the denominator was for instance (m-3) then we would need to multiply by (m-1)(m+2).
- We need to multiply by the ratio of Missing-factors Ł

$$\frac{\text{missing} - \text{factor(s)}}{\text{missing} - \text{factor(s)}}$$

which is

$$\frac{(m-1)(m+2)}{(m-1)(m+2)}$$

or

$$\frac{(m+2)(m-3)}{(m+2)(m-3)}$$

or

$$\frac{(m-1)(m-3)}{(m-1)(m-3)}$$

depending on the denominator which will result in the denominators being the same which is

$$(m-1)(m+2)(m-3)$$

- This is algebraically the same as multiplying by 1 since the same
  numerator and denominator equals one (1) one and will not change the value of the fraction when multiplying by the equivalent of one.

Complex Equation to be solved

$$\frac{\frac{1}{m-1}+\frac{2}{m+2}}{\frac{2}{m+2}-\frac{1}{m-3}}$$

Use sets of Missing-factors used to create the
LCD (lowest common denominator)

The Lowest common denominator in this example is:

$$(\mathrm{m}-1)(\mathrm{m}+2)(\mathrm{m}-3)$$

Note: To take the first example of

$$\frac{(m-1)(m+2)}{(m-1)(m+2)}$$

Note That:

$$\frac{\text{Missing} - \text{factors}}{\text{Missing} - \text{factors}} = \frac{(m-1)(m+2)}{(m-1)(m+2)} = 1 = \text{ a unity value}$$

Therefore, multiplying by a unity value (one) does not change the value of an expression although the appearance will change. Also, since this is one of the missing-factors it will create the factor

$$(m-1)\,(m+2)\,(m-3)$$

in the denominator of one of the of the sets. Similarly, we also multiply by the

$$\frac{\text{Missing} - \text{factors}}{\text{Missing} - \text{factors}}$$

to create the denominator in the correct factors (lowest common denominator).

Note: we also multiply the missing factor(s) in the numerator creating a new correct expression.

Complex Fraction to be solved

$$\frac{\frac{1}{m-1}+\frac{2}{m+2}}{\frac{2}{m+2}-\frac{1}{m-3}}$$

Multiply by

$$\frac{\text{Missing}-\text{factors}}{\text{Missing}-\text{factors}}$$

$$\frac{\frac{1}{m-1}\frac{(m+2)(m-3)}{(m+2)(m-3)}+\frac{2}{m+2}\frac{(m-1)(m-3)}{(m-1)(m-3)}}{\frac{2}{m+2}\frac{(m-1)(m-3)}{(m-1)(m-3)}-\frac{1}{m-3}\frac{(m-1)(m+2)}{(m-1)(m+2)}}$$

$$\frac{\frac{1}{m-1}\frac{(m^2+2m-3m-6)}{(m+2)(m-3)}+\frac{2}{m+2}\frac{(m^2-m-3m+3)}{(m-1)(m-3)}}{\frac{2}{m+2}\frac{(m^2-m-3m+3)}{(m-1)(m-3)}-\frac{1}{m-3}\frac{(m^2-m+2m-2)}{(m-1)(m+2)}}$$

Now that when the denominators are the same, we keep the denominators and add or subtract the terms in the numerator

$$\frac{\frac{3m^2-9m}{(m-1)(m+2)(m-3)}}{\frac{1}{m+2}\frac{(2m^2-2m-6m+6)}{(m-1)(m-3)}-\frac{1}{m-3}\frac{(m^2-m+2m-2)}{(m-1)(m+2)}}$$

Now do the same for the denominator in the numerator, and the denominator in the denominator.

$$\frac{\frac{3m^2-9m}{(m-1)(m+2)(m-3)}}{\frac{2m^2-2m-6m+6-m^2-m+2}{(m+2)(m-1)(m-3)}} = \frac{\frac{3m^2-9m}{(m-1)(m+2)(m-3)}}{\frac{m^2-9m+8}{(m+2)(m-1)(m-3)}}$$

To resolve the large fraction bar, invert the denominator and multiply times the numerator

$$\frac{3m^2-9m}{(m-1)(m+2)(m-3)} * \frac{(m+2)(m-1)(m-3)}{m^2-9m+8} =$$

$$\frac{3m^2-9m}{(m-1)(m+2)(m-3)} * \frac{(m+2)(m-1)(m-3)}{m^2-9m+8} =$$

$$\frac{3m^2-9m}{m^2-9m+8}$$

$$\frac{3m^2-9m}{(m-1)(m-8)}\text{ans}$$

Chapter 43

Simpler method of solving complex fractions

A simpler method of solving a complex fraction is to find the lowest common
denominator separately for the numerator for the fraction and the denominator of the fraction.

Solve this example.

$$\frac{\frac{1}{m-1}+\frac{2}{m+2}}{\frac{2}{m+2}-\frac{1}{m-3}}$$

$$\text{The } \frac{\text{missing}-\text{factor(s)}}{\text{missing}-\text{factor(s)}} \text{ are } \frac{m+2}{m+2} \text{ or } \frac{m-1}{m-1}$$

for the numerator,

and

$$\text{The } \frac{\text{missing} - \text{factor(s)}}{\text{missing} - \text{factor(s)}} \text{ are } \frac{m+2}{m+2} \text{ or } \frac{m-3}{m-3}$$

for the denominator.

Multiply by

$$\frac{\text{missing factor(s)}}{\text{missing factor(s)}}$$

$$\frac{\frac{1}{m-1}\frac{(m+2)}{(m+2)} + \frac{2}{m+2}\frac{(m-1)}{(m-1)}}{\frac{2}{m+2}\frac{(m-3)}{(m-3)} - \frac{1}{m-3}\frac{(m+2)}{(m+2)}}$$

Collect terms in the numerator and denominator fractions

Now add the numerators and keep the denominator

$$\frac{\frac{m+2}{m-1(m+2)} + \frac{2m-2}{m+2(m-1)}}{\frac{2m-6}{(m+2(m-3)} - \frac{(m+2)}{(m-2)(m-3)}}$$

$$\frac{\frac{m+2+2m-2}{(m+2)(m-1)}}{\frac{(2m-6)-(m+2)}{(m-2)(m-3)}} = \frac{\frac{3m}{(m+2)(m-1)}}{\frac{m-8}{(m-2)(m-3)}}$$

Invert the denominator and multiply

$$\frac{3m}{(m+2)\,(m+1)} * \frac{(m-3)\,(m+2)}{m-8}$$

$$\frac{3m}{(m+1)} * \frac{(m-3)}{m-8} = \frac{3m^2-9m}{(m+1)(m-8)} \quad \text{Answer}$$

Chapter 44

Rationalizing-the-denominator

The radical expression $\sqrt{\frac{3}{2}} = \frac{\sqrt{3}}{\sqrt{2}}$ is rationalized

by multiplying the radical expression by the unity-factor $\frac{\sqrt{2}}{\sqrt{2}}$.

$$\frac{\sqrt{3}}{\sqrt{2}} * \frac{\sqrt{2}}{\sqrt{2}} = \frac{\sqrt{2}\sqrt{3}}{\sqrt{2}\sqrt{2}} = \frac{\sqrt{6}}{2},$$

which satisfies the requirement that in a solved radical expression, the solved denominator cannot contain a radical.

44.1 Exp.2 Rationalizing Radicals denominators.

Using the equalization principle in the previous example:

$$\frac{\sqrt{2}}{\sqrt{2}} = 1$$

is a unity factor, and multiplying by a unity-factor does not change the value of the expression. Multiplying by a unity-factor produces an equivalent form of the expression. Therefore,

$$\sqrt{\frac{3}{2}} = \frac{\sqrt{3}}{\sqrt{2}} = \frac{\sqrt{3}}{\sqrt{2}} * \frac{\sqrt{2}}{\sqrt{2}} = \frac{\sqrt{6}}{2}$$

is a valid process for rationalizing the denominator.

44.2 An example of multiplying radicals

$$\frac{\sqrt{6x^2y}}{3}$$

Separate the x^2 into a separate radical factor

$$\text{Resolve } \sqrt{x^2} = x$$

$$\frac{\sqrt{x^2}\sqrt{6y}}{3} = \frac{x\sqrt{6y}}{3} \quad \text{Answer}$$

44.3 Rationalizing denominators with a cubic root.

$$\frac{\sqrt[3]{3}}{\sqrt[3]{2}} = \frac{\sqrt[3]{3}}{\sqrt[3]{2}} * \frac{\sqrt[3]{4}}{\sqrt[3]{4}} = \frac{\sqrt[3]{3*4}}{\sqrt[3]{2*4}} = \frac{\sqrt[3]{3*4}}{\sqrt[3]{8}} = \frac{\sqrt[3]{12}}{\sqrt[3]{8}} = \frac{\sqrt[3]{12}}{2}$$

Multiply the fraction by a unity factor that will make the denominator a perfect cube

$$\frac{\sqrt[3]{4}}{\sqrt[3]{4}}$$

and take the cube root of the denominator to simplify.

We can also express the result as

$$\frac{\sqrt[3]{12}}{2} = \frac{1}{2}\sqrt[3]{12}$$

44.4 Multiply an integer times a radical expression

Simplify $5(\sqrt{8} - \sqrt{32})$
First factor out the perfect squares $5(\sqrt{8}-\sqrt{32})$ $\sqrt{4}$ in the first ex
remaining in both expressions producing a like radical in both term

$5\left(\sqrt{(4)(2)} - \sqrt{(16)(2)}\right)$
$5\left(2\sqrt{2} - 4\sqrt{2}\right) = 10\sqrt{2} - 20\sqrt{2} = -10\sqrt{2}$

44.5 Rationalizing an expression with a radical in the denominator

$$\frac{5}{\sqrt{3}} = \frac{5}{\sqrt{3}} * \frac{\sqrt{3}}{\sqrt{3}} = \frac{5\sqrt{3}}{3} = \frac{5}{3}\sqrt{3}$$

$$\frac{\sqrt{3}}{\sqrt{2}} = \left(\frac{\sqrt{3}}{\sqrt{2}} * \frac{\sqrt{2}}{\sqrt{2}}\right) = \frac{\sqrt{6}}{2} = \frac{1}{2}\sqrt{6}$$

- Principals: Multiplying by the $\frac{\sqrt{2}}{\sqrt{2}}$ is algebraically the same as multiplying by one, since $\frac{\sqrt{2}}{\sqrt{2}}$ is a unity factor,

- Multiplying by
$$\frac{Numerator1}{\sqrt{2}} * \frac{Numerator2}{\sqrt{2}} = \frac{New\ Numerator\,(1*2)}{2}$$
resulting with 2 in the denominator. Since $\left(\sqrt{2}\right)^2 = \left(2^{\frac{1}{2}}\right)^2 = (2)^1 = 2$ in the denominator.
- Multiplying by a unity factor does not change the value of the expression and creates an equivalent form of the expression which can be now solved.
- In a solved radical expression, the denominator cannot contain a radical.

44.6 Exp.2 Rationalizing Radicals denominators.

44.7 Rationalizing denominators with a cubic root.

$$\frac{\sqrt[3]{3}}{\sqrt[3]{2}} = \frac{\sqrt[3]{3}}{\sqrt[3]{2}} * \frac{\sqrt[3]{2^2}}{\sqrt[3]{2^2}} = \frac{\sqrt[3]{3*2^2}}{\sqrt[3]{2*2^2}} = \frac{\sqrt[3]{3*2^2}}{\sqrt[3]{2^3}} = \frac{\sqrt[3]{12}}{2} = \frac{1}{2}\sqrt[3]{12}$$

Multiply the fraction by a unity factor that will make the denominator a perfect cube
$$\frac{\sqrt[3]{2^2}}{\sqrt[3]{2^2}}$$
and take the cube root of the denominator to simplify.

We can express the result as
$$\frac{\sqrt[3]{12}}{2} = \frac{1}{2}\sqrt[3]{12}$$

Chapter 45

Definition of a Radical

Radical $= \sqrt[n]{a}$

Where n equals the index.

Where $\sqrt{}$ equals the radical sign and

Where $\sqrt[n]{a}$ equals the entire radical

Where a is the radicand under the radical sign.

45.1 Examples of radicals

Radicals are expressions under a root sign.

$\sqrt{x}$ or $\sqrt{x+2}$ are two different examples.

$\sqrt[2]{\text{x}}$ is expressed as $\sqrt{\text{x}}$ where the second root is implied

Higher roots than 2 such as 3, in $\sqrt[3]{x}$ are always specified.

Chapter 46

Adding-and-subtracting-radicals

46.1 Add or subtract Radicals with the same root

If a radical expression contains sums or differences of Radicals that are the same under the root sign, add the coefficients and keep the root. I.E. $3\sqrt{2} + 4\sqrt{2} = 7\sqrt{2}$

46.2 Resolving Radicals

After simplifying by extracting, the perfect squares, the expressions both have the same remaining radical $\sqrt{2}$.

Simplify

$5 \bullet (\sqrt{8} - \sqrt{32})$

Start by simplifying

$5(\sqrt{8} - \sqrt{32})$

$\sqrt{5} \bullet (\sqrt{8} - \sqrt{32})$

$5\left(\sqrt{(4)\,(2)} - \sqrt{(16)\,(2)}\right)$

$5\left(2\sqrt{2} - 4\sqrt{2}\right) = 10\sqrt{2} - 20\sqrt{2} = -10\sqrt{2}$

Chapter 47

Multiplying-and-dividing-roots

47.1 Product rule for radicals

$\sqrt{x} \bullet \sqrt{y} = \sqrt{xy}$

$\sqrt{3} \bullet \sqrt{2} = \sqrt{6}$

$\sqrt{5} \bullet \sqrt{x} = \sqrt{5x}$

Solve the following:

$$\left(\sqrt{3} + 2\sqrt{5}\right)\left(\sqrt{3} - 4\sqrt{5}\right)$$

Solve by using FOIL (First, Outer, Inner, Last)

$First$	$Outer$	$Inner$	$Last$
$\sqrt{3} \bullet \sqrt{3}$	$\sqrt{3} \bullet \left(-4\sqrt{5}\right)$	$2\sqrt{5} \bullet \sqrt{3}$	$2\sqrt{5} \bullet \left(-4\sqrt{5}\right)$

$$3 \quad -4\sqrt{15} \quad + \quad 2\sqrt{15} \quad -8\,(5)$$

Combined middle terms

$$-37 - 2\sqrt{15} \qquad \text{Answer}$$

47.2 Special products rule application

Special products rules are found on page 145.

The special products rule for an algebraic term,squared.

$$(a+b)^2 = a^2 + 2ab + b^2$$
$$(a-b)^2 = a^2 - 2ab + b^2$$

Apply the special products rule to solve $\left(\sqrt{10}-7\right)^2$

$$\left(\sqrt{10}-7\right)^2; \quad \text{Let } a = \sqrt{10} \quad \text{and} \quad b = 7$$

$$\left(\sqrt{10}-7\right)^2 = \left(\sqrt{10}\right)^2 - 2 \bullet \sqrt{10} \bullet 7 + (7)^2$$

$$10 - 14\sqrt{10} + 49$$
$$59 - 14\sqrt{10} \quad \text{answer}$$

47.3 Reducing binomial multiplication with radicals

The following is an example reducing binomial multiplication radicals using FOIL method. Solve the following:

$$\left(\sqrt{10}-\sqrt{5}\right) \bullet \left(\sqrt{5}+\sqrt{20}\right)$$

General form:

$$(a-b) \bullet (b+c)$$

$$(a-b) \bullet (b+c) = ab + bc - b^2 - bc$$

$$\left(\sqrt{10} \bullet \sqrt{5}\right) + \left(\sqrt{10} \bullet \sqrt{20}\right) - \left(\sqrt{5}\right)^2 - \left(\sqrt{5} \bullet \sqrt{20}\right)$$

$$\sqrt{50} + \sqrt{200} - 5 - \sqrt{100}$$

Reduce the radicals by factoring out the perfect squares.

$$\sqrt{25 \bullet 2} + \sqrt{100 \bullet 2} - 5 - \sqrt{100}$$

$$5\sqrt{2} + 10\sqrt{2} - 5 - 10$$

Collect like terms.

$$15\sqrt{2} - 15 \quad \text{answer}$$

47.4 Complete the solution by rationalizing the denominator

$$\sqrt{\frac{7}{13}} \bullet \sqrt{\frac{13}{3}} = \sqrt{\frac{7 \bullet 13}{13 \bullet 3}} = \sqrt{\frac{7}{3}}$$

Rationalize the denominator by multiplying by

$$\sqrt{\frac{3}{3}}$$

$$= \sqrt{\frac{7}{3}} \bullet \sqrt{\frac{3}{3}} = \sqrt{\frac{7 \bullet 3}{9}} = \frac{\sqrt{21}}{3} \text{ or } \frac{1}{3}\sqrt{21} \text{ answer}$$

Chapter 48

Resolving Radicals

48.1 Factoring out a perfect square Evample 1

If a radicand has a factor that is a perfect square (in the case of square roots) express the radical as a product of the perfect square times the remaining factor.

$\sqrt{20} = \sqrt{5 * 4} = \sqrt{4} * \sqrt{5} = 2\sqrt{5}$

$\sqrt[3]{16} = \sqrt[3]{8 * 2} = \sqrt[3]{8} * \sqrt[3]{2} = 2\sqrt[3]{2}$

48.2 Simplify a Radical Expression Example 2

Simplify $5 * (\sqrt{8} - \sqrt{32})$
Start by simplifying inside the parenthesis
$5(\sqrt{8} - \sqrt{32})$
$\sqrt{8}$ and $\sqrt{32}$ can be reduced by separating out the perfect squares.
$\sqrt{5} * (\sqrt{8} - \sqrt{32})$
$5\left(\sqrt{(4)(2)} - \sqrt{(16)(2)}\right)$
$5\left(2\sqrt{2} - 4\sqrt{2}\right) =$
$10\sqrt{2} - 20\sqrt{2} = -10\sqrt{2}$

48.3 Simplify a Radical Expression Example 3

$$\frac{5}{\sqrt{3}} = \frac{5}{\sqrt{3}} * \frac{\sqrt{3}}{\sqrt{3}} = \frac{5\sqrt{3}}{3} = \frac{5}{3}\sqrt{3}$$

48.4 Unity Factor Explanation

$$\frac{\sqrt{3}}{\sqrt{3}}$$

is a unity factor and multiplying by a unity factor does not change the value of the expression, only produces an equivalent form of the expression. $\frac{5}{\sqrt{3}}$ is rationalized by multiplying the fraction by $\frac{\sqrt{3}}{\sqrt{3}}$ equals 1, a unity multiplier $\frac{5}{\sqrt{3}}$ This yields $5\sqrt{3}$and 3 in the denominator which satisfies the requirement that In a solved radical expression the denominator also $\frac{\sqrt{2}}{\sqrt{2}}$ is a unity factor.

48.5 Simplify a Cubics Example 4

Using the same rules as the previous example Multiply the fraction by a unity factor that will make the denominator a perfect cube

$$\frac{\sqrt[3]{2^2}}{\sqrt[3]{2^2}}$$

and take the cube root of the denominator to simplify. We can also express the result as

$$\frac{\sqrt[3]{3 * 2^2}}{\sqrt[3]{2 * 2^2}} = \frac{\sqrt[3]{3 * 2^2}}{\sqrt[3]{2^3}} = \frac{\sqrt[3]{12}}{2} = \frac{1}{2}\sqrt[3]{12}$$

48.6 Unity Factor

The $[\frac{\sqrt{3}}{\sqrt{3}}]$ is rationalized by multiplying the fraction by $[\frac{\sqrt{3}}{\sqrt{3}}]$ equals 1, a unity multiplier $\frac{5}{\sqrt{3}}$. This yields $\frac{5\sqrt{3}}{3}$ which satisfies the requirement that, In a solved radical expression the denominator cannot contain a radical. This can be re-wrtten as $\frac{1}{3} * 5\sqrt{3}$

48.7 Rationalizing the denominator Example 5

$$\sqrt{\frac{2x^2y}{3}} = \left(\sqrt{\frac{2x^2y}{3}} * \frac{\sqrt{3}}{\sqrt{3}}\right) = \frac{\sqrt{6x^2y}}{3}$$

$$\frac{\sqrt{x^2}\sqrt{6y}}{3} = \frac{x\sqrt{6y}}{3}$$

48.8 Perfect squares reduction Example 6

$$\sqrt{\frac{169}{9}} = \frac{13}{3}$$

Since both numerators are perfect squares take the square root of both terms.

Appendix A

Review of Quadratics

A.1 Review of a quadratic equation.

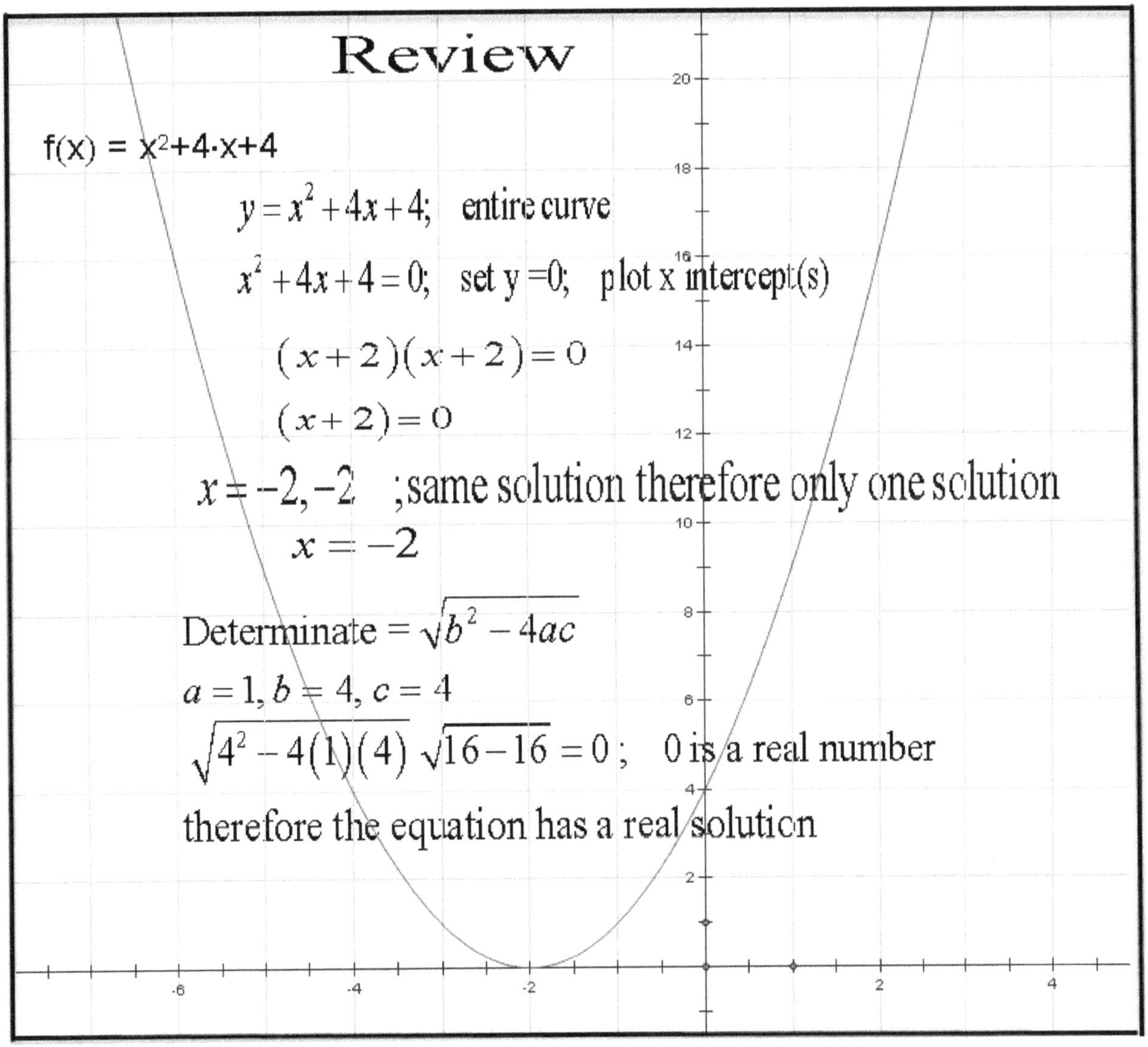

Solution$\{x, y\} = \{-2, 0\}$

Figure A.1: review of quadratics and Graph

Appendix B

Algebra rules apples other higher-level math

B.1 Trigonometry Discussion

In trig we layer on rules of trigonometric functions also.

In Trigonometry, functions of trigonometry replace some of the variables but the rules for manipulating variables are the same .

i.e. We use $\sin(x)$ $\cos(x)$ or $\tan(x)$ as variables as well as x and y and z when needed to solve the problem.

B.2 Calculus Discussion

In calculus the some of the variables are constantly changing according the rules of calculus.

Variables are time, speed, distance and acceleration,

The variables would be t,s,v,a, in acceleration problems.

Some of the variables will continually change during the duration of the problem

Appendix C

Distance-Formulas

C.1 Formula for distance of an object projected vertically: Physics Problem

This formula is used in a vertical problem found on page 115

Vertical Distance: $= V_{y0}t - \frac{1}{2}gt^2$

Acceleration due to gravity $g = 32\frac{ft}{\sec^2}$

$y = f\ (t)$y is dependent on t for its solution

Change the order of terms to the format of a quadratic equation

Replace y with f(t). Note: f(t) states that y is dependent on t for its solution

Original equation: $y = V_{y0}t - \frac{1}{2}gt^2$

$H\ =\ s(t)$ equals height of ball

at any time, and is dependent on

the time elapsed.

$H = s(t) = -\frac{1}{2}gt^2 + V_{y0}t\ +\ H_0$ feet

Where $H_0\ =\ $ initial height above ground

Acceleration due to gravity $32\frac{ft}{\sec^2}$

This fits into the general format for a Quadratic Equation

$y = f(x) = ax^2 + bx + c$

$s(t) = -16t^2 + 16t + 192$

$y = f(x) = -\frac{1}{2}\left(32\frac{ft}{\sec^2}\right)t^2 + V_{y0}t \ + \ 0 \text{ feet}$

$y = f(x) = -16\ t^2 + V_{y0}t \ + \ 0 \text{ feet}$

$f(x) = -16\ t^2 + V_{y0}t \ + \ 0 \text{ feet}$

In the example given earlier $V_{y0}t$ the initial velocity was $16\frac{ft}{\sec}$

The initial Height was 192 feet

Substituting these values

$s(t) = -16t^2 + 16t + 192$

$s(t) = -\frac{1}{2}gt^2 + V_{y0}t \ + \ x \text{ feet}$

This formula is used in a vertical problem found on page 115

Thrown ball from a roof top: Quadratic Equation

t = 3 seconds to reach 96 feet

t = 4 seconds to reach 0 feet

Where 0 feet is at ground level.

Appendix D

Mathematics use in Electrical Engineering

Algebra is used in Mechanical Engineering,Civil Engineering, Chemistry and all other sciences.

D.1 Electrical Engineering

Both in Electrical Engineering and Practical Math Applications used in solving Electrical Design Problems. Were used every day of my projects. It was our task to debug new designs and design test equipment and make both the designs and test equipment work before the design could be approved.

D.2 Proofs as Student handouts to add to explanation of principles.

We used a lot of practical mathematics putting to use the Math and physics and electrical theory that we learned in engineering school. My teaching was to help understand the theory to make tests and later career problem solving easier and more fun. To that end I started writing proofs that were used as handouts to the students.

D.3 Proofs

If a student understands how to derive a formula the student can adapt the formula for solving similar math problems.

Note: The imaginary coordinate system is used in electrical work where $-i$ values are capacitance and $+i$ values are inductance and resistance is plotted along the x-axis.

In Electrical Engineering the curve showing a transistor or diode curve also tells us in a v versus vi curve indicates what setting will give us the most stability. This is very important when designing an amplifier or a computer.

In Mechanical Engineering the elasticity the linear portion of the curve show how much stress can be applied to a metal, wood or plastic device. Too much stress will cause a breaking point potentially. This used in designing a new part.

Appendix E

Approximate numbers

E.1 Number description

- Numbers can be Integers or Approximate Numbers
- Integers are used in counting.
- In Engineering Approximate numbers are important.
- Every measurement is approximate and depends on the accuracy of the instrument taking the measurement.
- This is true in any type of science.

E.2 Critical factors of approximate numbers

- Precision
- Accuracy

E.3 Definition approximate numbers

- Precision depends on number of places after the decimal point.
- Accuracy depends on number of significant digits. (see chart)

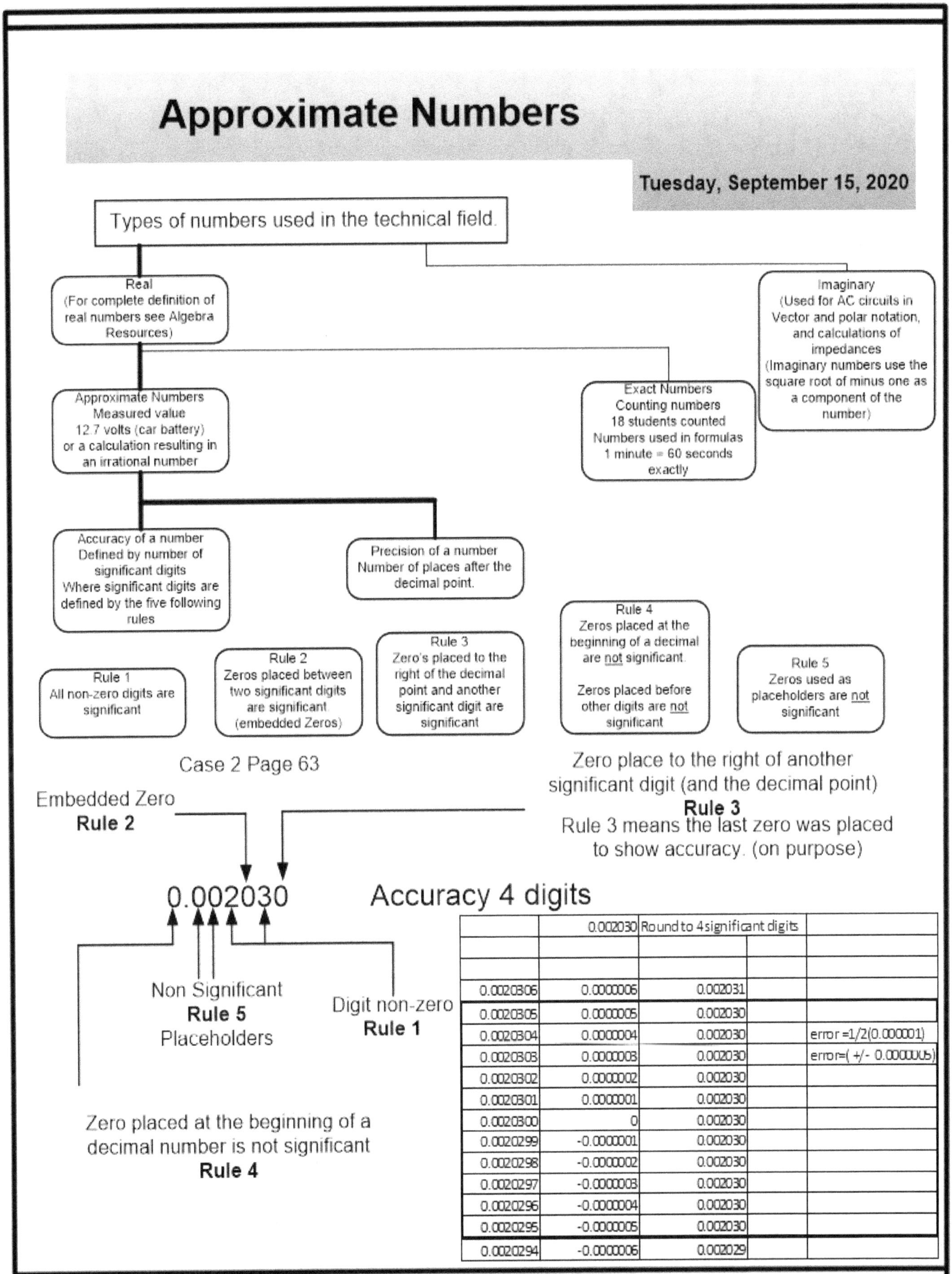

	0.002030	Round to 4 significant digits		
0.0020306	0.0000006	0.002031		
0.0020305	0.0000005	0.002030		
0.0020304	0.0000004	0.002030		error =1/2(0.000001)
0.0020303	0.0000003	0.002030		error=(+/- 0.0000005)
0.0020302	0.0000002	0.002030		
0.0020301	0.0000001	0.002030		
0.0020300	0	0.002030		
0.0020299	-0.0000001	0.002030		
0.0020298	-0.0000002	0.002030		
0.0020297	-0.0000003	0.002030		
0.0020296	-0.0000004	0.002030		
0.0020295	-0.0000005	0.002030		
0.0020294	-0.0000006	0.002029		

Figure E.1: Approximation of numbers

Appendix F

Math used in Physics

F.1 Ramp forces Diagrams

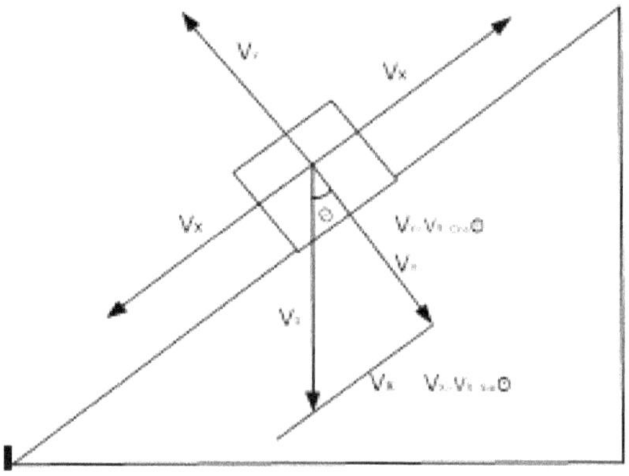

This outline shows the Physics Force Vectors

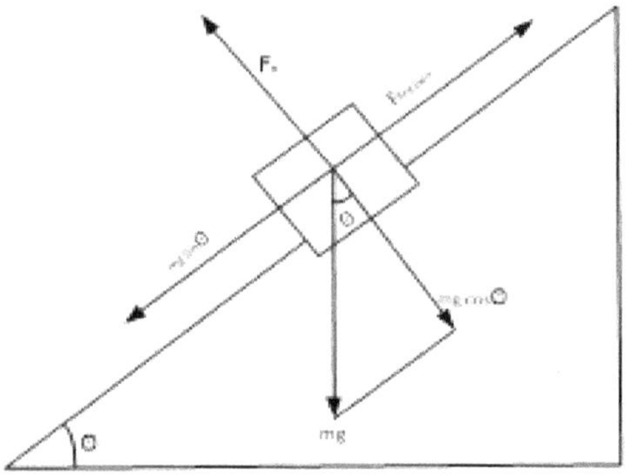

Trigonometry

Geometry

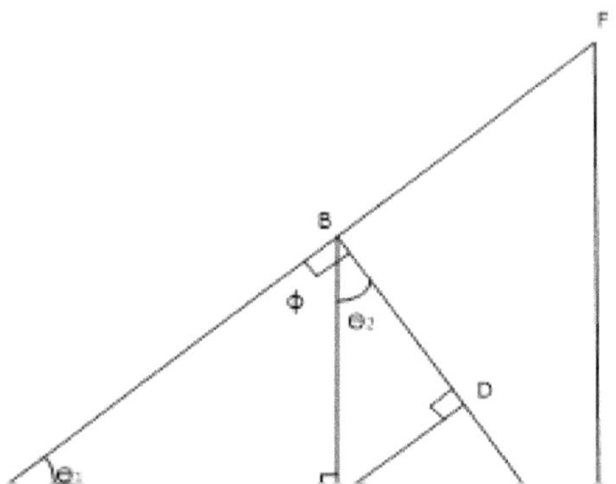

Forces of a block on a ramp. Where a vector has magnitude and direction. Newton's 3rd law For every action there is an equal and opposite reaction. The interaction of the block on the ramp creates a set of interactive forces in both the vertical and horizontal direction. The block also has a force of gravity in the y direction

Trigonometry

$$\sin\theta = \frac{\text{Opp}}{\text{Hyp}} = \frac{F_{gx}}{F}; \textit{therefore } F_{gx} = F\sin\theta$$

$$\cos\theta = \frac{\text{Adj}}{\text{Hyp}} = \frac{F_{gy}}{F}; \textit{therefore } F_{gy} = F\cos\theta$$

Geometry

Prove that the ramp angle CAB is congruent to angle CBD.
AB $\perp$ BD and CD $\perp$ BD and AC $\perp$ BC therefore forming a series of right angle triangles namely ▲ABC ▲ CBD.

Where CAB= Θ_1 and CBE= Θ_2 and ABC= Φ.

1) $\Theta_1+\Phi=90^0$ therefore $\Theta_1=90^0-\Phi$
2) $\Theta_2+\Phi=90^0$ therefore $\Theta_2=90^0-\Phi$
   Statmt 1) sum of a rt $\Delta=180^0$
   Statmt 2) sum complementary angles equals 90^0
3) Then $\Theta_1=\Theta_2$ and are congruent

F.2 convert-units.png

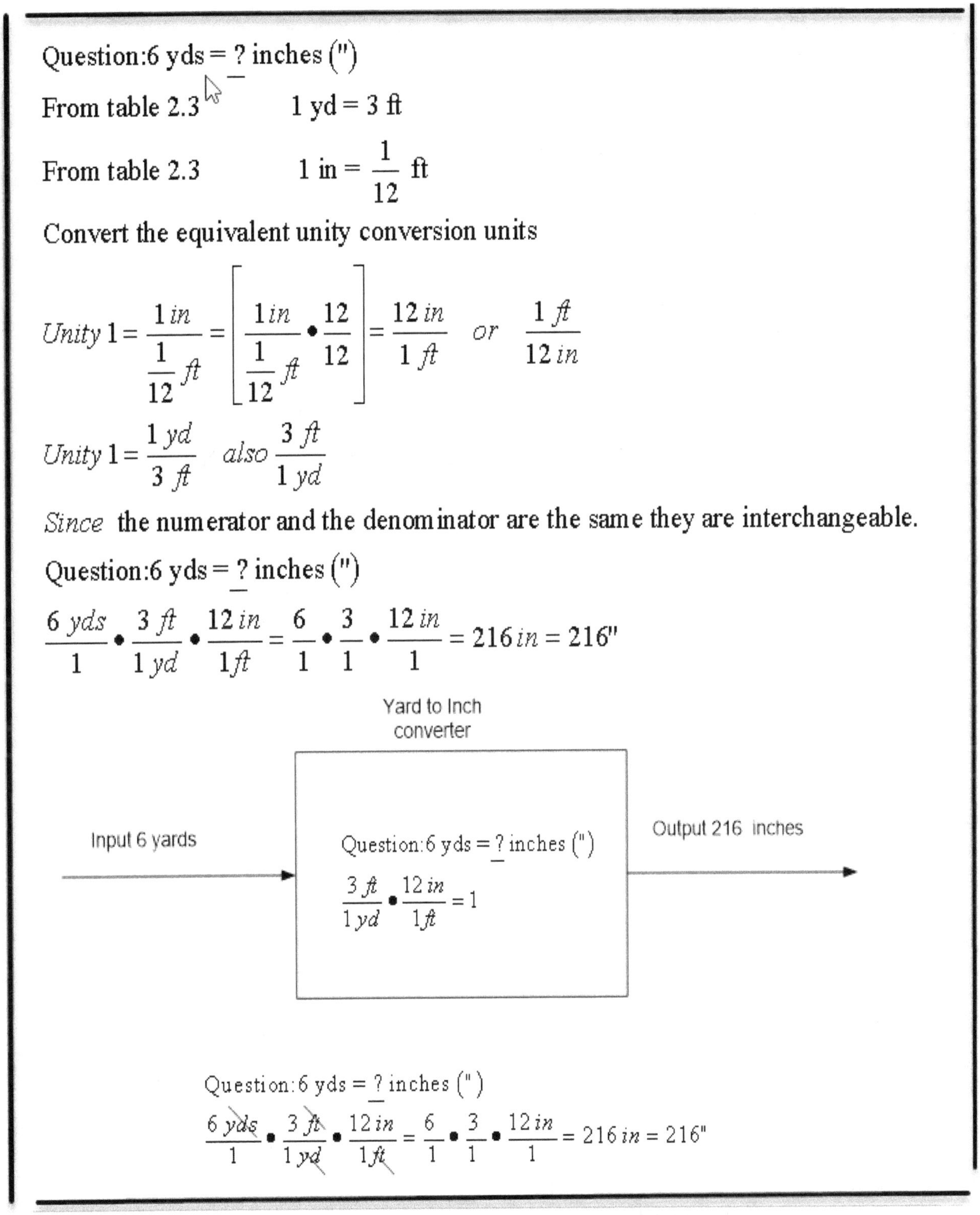

Figure F.2: convert-units.png

Appendix G

Calculus:Approximate-vs-Integral area under a curve

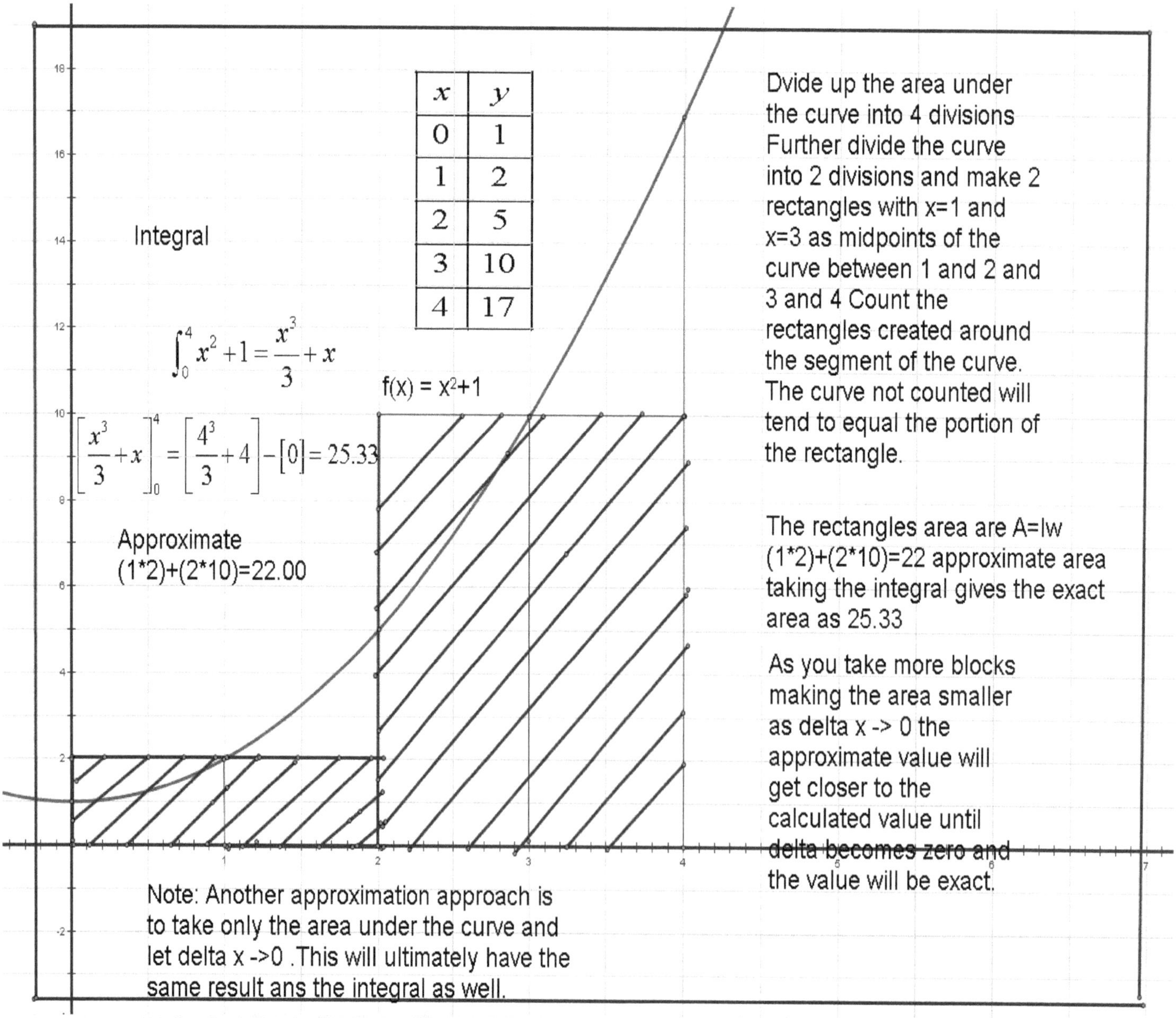

Figure G.1: approximate-vs-integral

G.1 secant

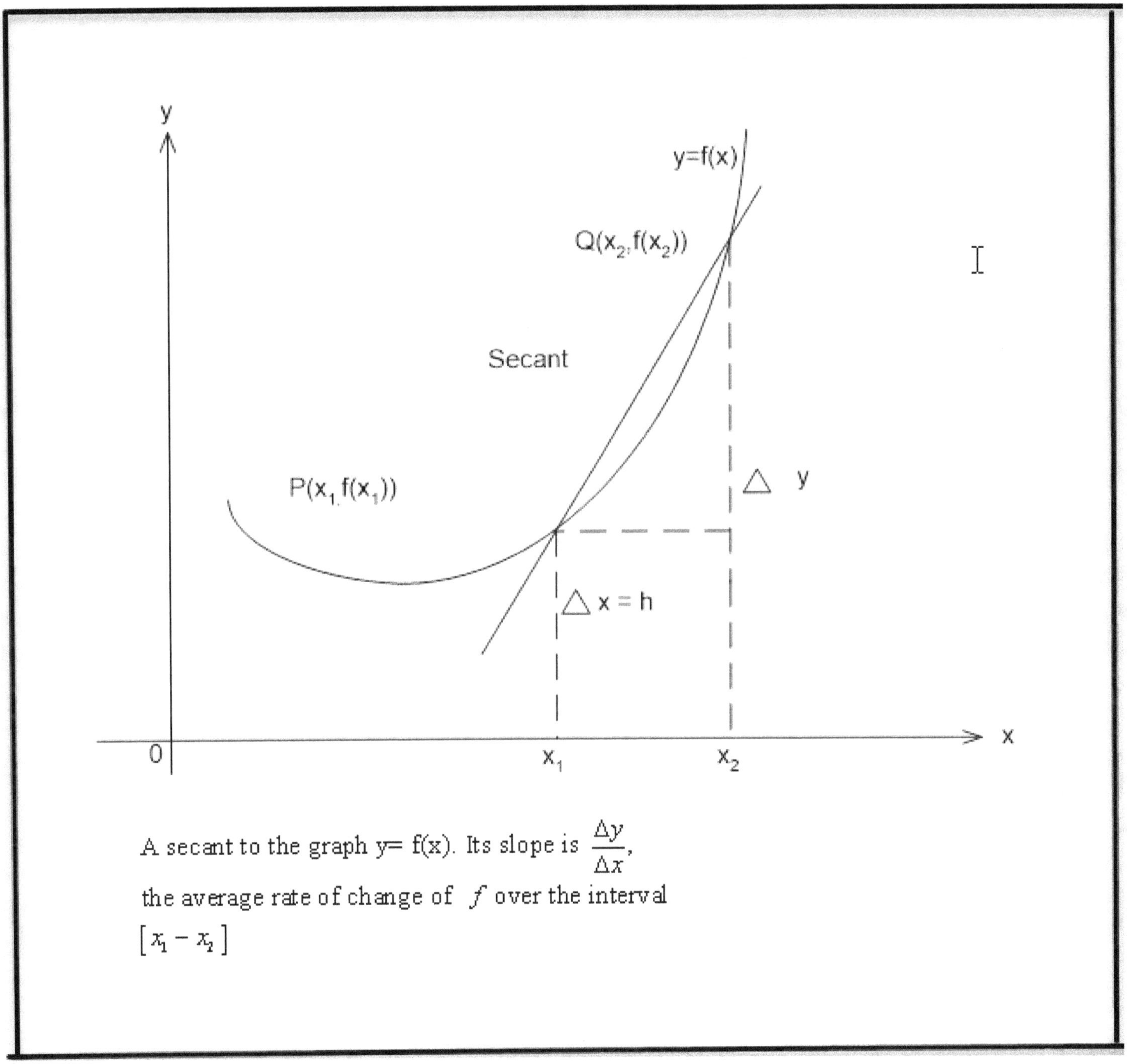

Figure G.2: secant

G.2 Limit of delta x

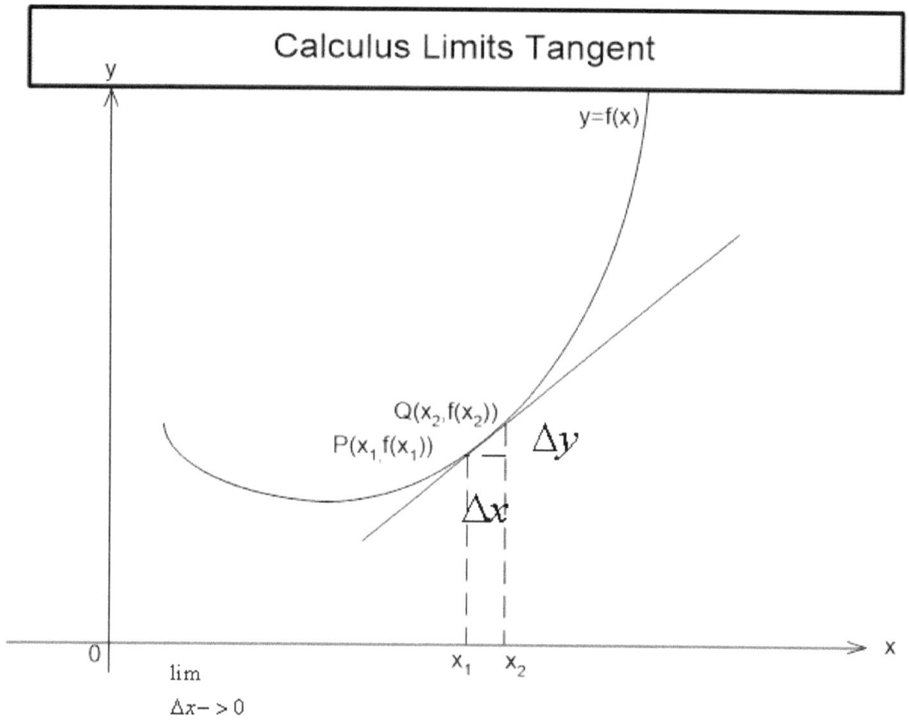

$\lim_{\Delta x \to 0}$

As Δx approaches zero (get much smaller) then

$\frac{\Delta y}{\Delta x}$ also approaches zero and x_2 moves towards x_1 $\left(\lim_{\Delta x \to 0}\right)$

Then the Secant line (which was the average slope of $P(x_1, f(x_1))$ and $Q(x_2, f(x_2))$ now approaches the slope of $P(x_1, f(x_1))$.

Graphically at Δx actually reaches *zero* P and Q become the same point and the slope is the tangent of P.

Figure G.3: Calculus Limits Tangent

G.3 finite-approximation.png

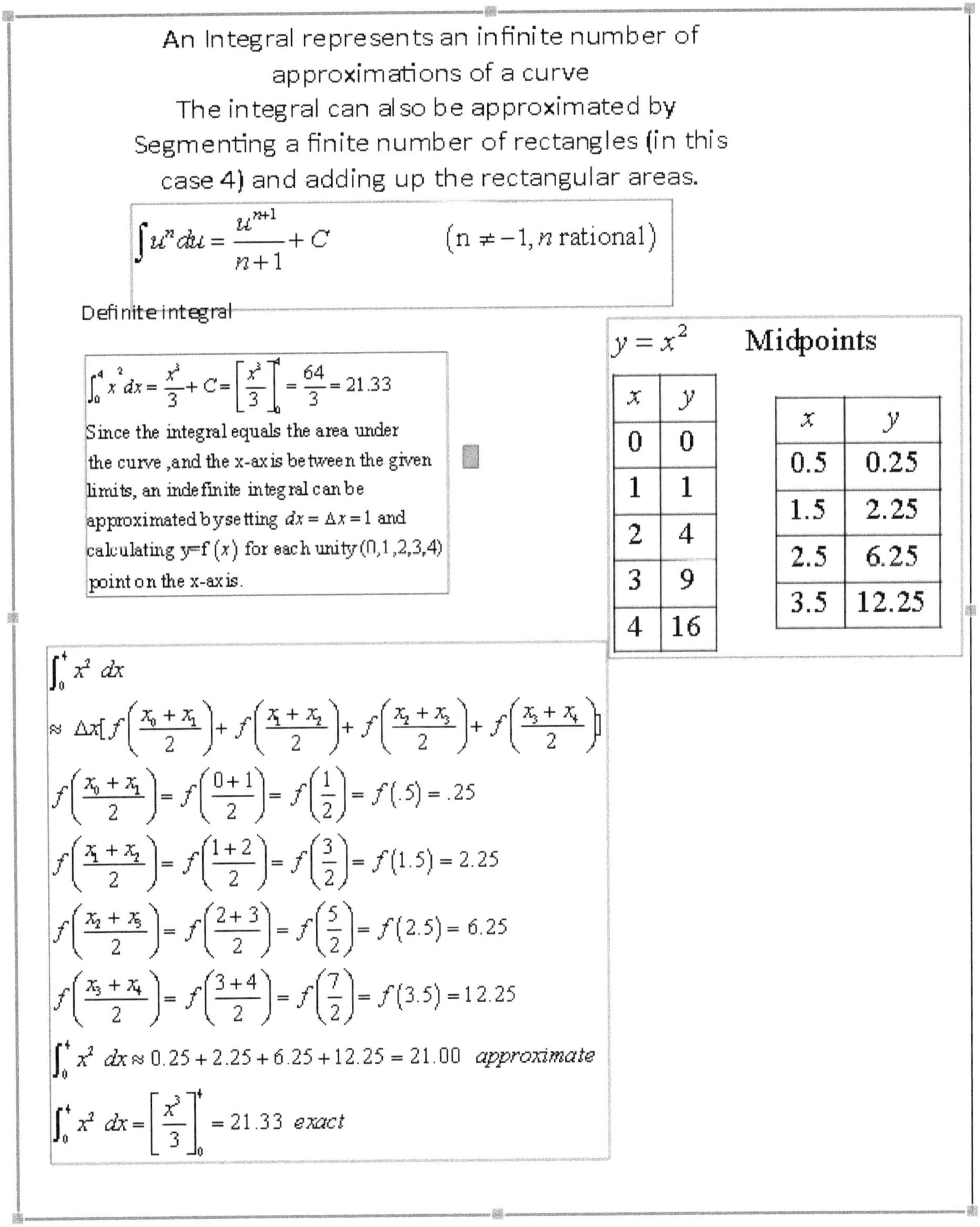

Figure G.4: finite-approximation.png

G.4 integral-by-parts

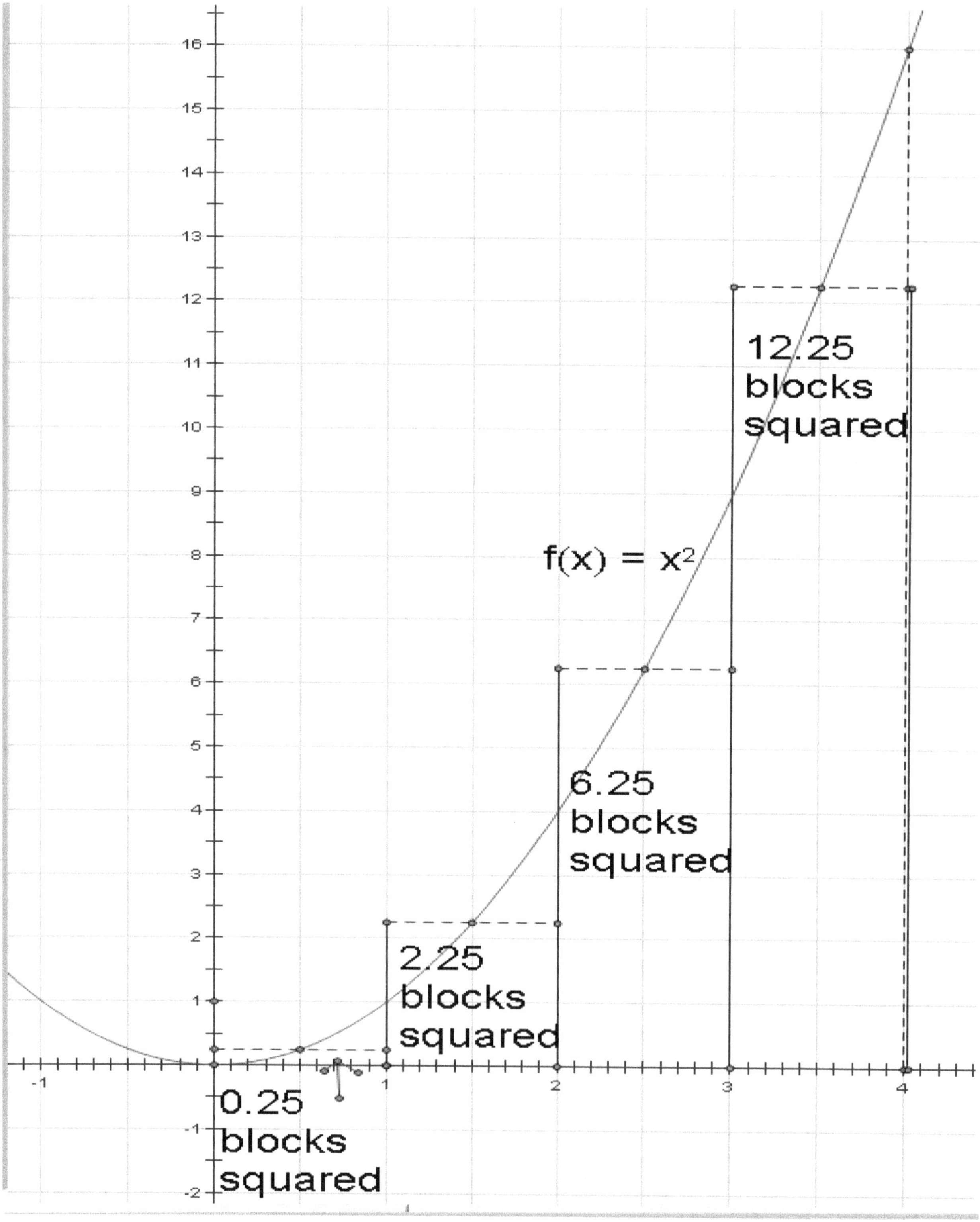

Figure G.5: integral-by-parts

www.ingramcontent.com/pod-product-compliance
Lightning Source LLC
LaVergne TN
LVHW060413220926
848484LV00015B/2531